Introduction to Intellectual Property Rights

Introduction to
Intellectual
Property Rights

HS Chawla
Professor and Head
Department of Genetics and Plant Breeding
GB Pant University of Agriculture and Technology
Pantnagar, India

Oxford & IBH Publishing Co. Pvt. Ltd.
New Delhi
(*A Unit of* CBS Publishers & Distributors Pvt Ltd)

CBS

CBS Publishers & Distributors Pvt Ltd

New Delhi • Bengaluru • Chennai • Kochi • Kolkata • Mumbai
Hyderabad • Jharkhand • Nagpur • Patna • Pune • Uttarakhand

Introduction to
**Intellectual
Property Rights**

ISBN-13 978-81-204-1797-7
ISBN-10 81-204-1797-6

© 2016, Copyright Reserved
CBS Reprint: 2020, 2022, 2023

OXFORD & IBH
New Delhi
(A Unit of CBS Publishers & Distributors Pvt Ltd)

Published by Satish Kumar Jain and produced by Varun Jain for

CBS Publishers & Distributors Pvt Ltd
4819/XI Prahlad Street, 24 Ansari Road, Daryaganj, New Delhi 110 002, India
Ph: 011-23289259, 23266861, 23266867   Website: www.cbspd.com
Fax: 011-23243014   e-mail: delhi@cbspd.com;
                                           cbspubs@airtelmail.in.

Corporate Office: 204 FIE, Industrial Area, Patparganj, Delhi 110 092, India
Ph: 011-4934 4934   Fax: 011-4934 4935   e-mail: publishing@cbspd.com;
                                              publicity@cbspd.com

Branches

- **Bengaluru:** Seema House 2975, 17th Cross, KR Road, Banasankari 2nd Stage, Bengaluru 560 070, Karnataka, India
  Ph: +91-80-26771678/79   Fax: +91-80-26771680   e-mail: bangalore@cbspd.com
- **Chennai:** 7, Subbaraya Street, Shenoy Nagar, Chennai 600 030, Tamil Nadu, India
  Ph: +91-44-26680620, 26681266   Fax: +91-44-42032115   e-mail: chennai@cbspd.com
- **Kochi:** 42/1325, 1326, Power House Road, Opp KSE3, Power House, Ernakulum Kochi 682 018, Kerala, India
  Ph: +91-484-4059061-65,67   Fax: +91-484-4059065   e-mail: kochi@cbspd.com
- **Kolkata:** 147, Hind Ceramics Compound, 1st Floor, Nilgunj Road, Belghoria, Kolkata-700056, West Bengal, India
  Ph: +91-9096713055/7798394118, 9836841399   e-mail: kolkata@cbspd.com
- **Lucknow:** Basement, Khushnuma Complex, 7 Meerabai Marg (Behind Jawahar Bhawan),Lucknow-226001, UP, India
  Ph: +0522-4000032   e-mail: tiwari.lucknow@cbspd.com
- **Mumbai:** PWD Shed, Gala no 25/26, Ramchandra Bhatt Marg, Next to JJ Hospital Gate no. 2, Opp. Union Bank of India, Noorbaug, Mumbai-400009, Maharashtra, India
  Ph: 022-66661880/89   e-mail: mumbai@cbspd.com

Representatives

- Hyderabad   0-9885175004
- Patna   0-9334159340
- Jharkhand   0-9811541605
- Pune   0-9623451994
- Nagpur   0-9421945513
- Uttarakhand 0-9716462459

Printed at Chaman Enterprises, Daryaganj, Delhi, India

Preface

Leadership of the world in the 21st century will increasingly be in the hands of those who create and harness knowledge. This century, known as the century of knowledge, is indeed the century of mind. A nation's ability to convert knowledge into wealth and social good through creativity and innovation will determine its future standing in the comity of nations. In today's world, confidence in the intellectual property system acts as a powerful stimulus to creativity and innovation. Intellectual Property Rights (IPRs) protection plays a key role in gaining an advantageous position in the competitive world for achieving economic growth. Understanding and appreciating the IPR, its nature and the rationale for protection is a prerequisite for comprehending its role in formulating strategies for enhancing competitiveness. This has become more pronounced in the globalized economy with the obligations in the field of IPR and their relevance in agriculture and biodiversity in the World Trade Organization (WTO) regime.

With the establishment of WTO, Member States are required to abide by its Agreement on Trade-Related Aspects of Intellectual Property Rights (TRIPs). It lays down minimum norms and standards for the protection of IPRs and more importantly, their effective enforcement, unlike any other international treaty in the field of intellectual property. As a follow up of the WTO Agreements, the legislative scenario concerning IPRs changed fast and by 2005 the country had revised and enacted various IPR laws to meet the national requirements vis-à-vis TRIPs Agreement. IPRs are important not only because India as a member is required to accede to the conditions of an international agreement but also because they offer possible mechanisms for stimulating research, enabling access to technology and promoting enterprise growth with an ultimate aim to benefit the community.

Intellectual Property Rights has been introduced as a subject across many academic institutions to meet course curricula requirement in the face of increasing globalization of trade and commerce. There is a growing demand for easier access to information and understanding of basic concepts through a text book on intellectual property. Several books are available for law students aspiring towards legal profession but not for students and teachers desirous of having an introductory insight into intellectual property rights, its protection and management.

The purpose of this book is to provide a basic knowledge and understanding of the subject on intellectual property rights in all spheres. This book assumes the reader has little prior knowledge about or expertise in intellectual property rights and thus it has been prepared primarily for introductory purposes. In each chapter a section on treaties, conventions or agreements relevant to particular form of IPRs and case studies has been included. The book has been divided into three parts. The first part deals with basics and different forms of IPRs, *viz.* Patents, Copyrights, Trade Marks, Industrial designs, Trade secrets, Geographical indications and Semi conductor integrated circuits layout design. Patenting in relation to biological material, biotechnology and bioinformatics has been dealt separately. Concept of valorization, licensing and commercialization of technology has been dealt as a chapter on technology management. The second part deals with plant variety protection, biodiversity and traditional knowledge with basic introduction to IPR protection of agriculture and biological resources. Plant variety protection mechanism has been discussed at both national and global level. The third part deals with biosafety and bioethical issues concerning biotechnological research and its IPR protection.

The expected outcome from this book is to create increased IPR awareness and literacy among students, scientists, innovators, farmers, and entrepreneurs and to fulfill the needs and demands of students for an introductory book on IPRs where courses have been designed but without any text book on the subject. It is my sincere hope that readers will gain insight from this publication and would encourage them to protect their knowledge driven inventions for higher productivity and competitiveness.

I am highly indebted to God for giving me the strength and courage to complete this book and hope his gracious blessings will continue to be showered on me to further improve this manuscript with your suggestions and relevant criticism. It will be most helpful during the preparation of subsequent revised edition.

I am grateful and indebted to my elders for giving me support and to my wife, children and grandchild Sifat Singh (Agam) for their continuous love and encouragement.

March, 2016   Dr. H.S. Chawla

Contents

PART 1

IPR: Different Forms

Intellectual Property Rights – An Introduction

1. Introduction and Meaning of Intellectual Property

In today's knowledge economy, it is mainly knowledge and its application that drives the progress of a nation. Nations acquire knowledge and vigorously protect it in order to get its maximum value through leveraging the application of this knowledge. Intellectual property rights (IPRs) gives a strong foundation and plays a key role in the knowledge economy by bringing the fruits of research and new knowledge to the society.

Intellectual property (IP) is a product of the mind and it belongs to you just like a house or mobile or car. House or mobile or car are tangible things which one can feel and see with naked eyes but what is in the mind of a person nobody knows unless and until this intangible thing in the mind of a person is produced in a tangible form of a product or a new creation like TV, mobile, laptop, book, song, etc. Intellectual property is intangible in contrast to tangible as real property (land) or physical property, which one can see, feel and use. With any type of property there are property rights. When IPs are expressed in a tangible form, they can also be protected. The rights given to the creators of intellectual property are referred as Intellectual Property Rights (IPRs) which have been created to protect the right of individuals to enjoy their creations and discoveries. In fact, IPRs can be traced back to the fourteenth century, when European monarchs granted proprietary rights to writers for their literary works.

In the world scenario French theorist Benzamin in 1818 first introduced idea of intellectual property related to his writings. However it was with establishment of World Intellectual Property Organization (WIPO) in 1967 that actual use of the term intellectual property (IP) began. The term intellectual property right (IPR) was used quite earlier in a patent case of Devoll v/s Brown (1885) where Justice Wood Berg observed that labours of the mind, productions and interests are as much a Man's own as the wheat he cultivates or the flocks he rears. Even as per Section 1 of the French Law of 1791, all new discoveries are the property of author and to assure the inventor the property and enjoyment of his discovery, a patent need to be delivered to him. Also the French author A. Nion quoted the term intellectual property in his book that came in 1846. Many such instances can be quoted from history to trace the origin of the idea of Intellectual Property Rights (IPRs). The initial conventions that recognized safeguarding of IPRs, were Paris Convention of 1883 for the protection of industrial property and the Berne Convention of 1886 for the protection of literary and artistic works.

The process of translating an idea or invention into a good or service that creates value or for which customers will pay is to be called as **innovation**. An idea must be replicable at an economical cost and must satisfy a specific need. In business, innovation often results when ideas are applied by the company in order to further satisfy the needs and expectations of the customers. Innovation is synonymous with risk-taking and organizations that create revolutionary products or technologies take on the greatest risk because they create new markets. Thus innovation is a change that unlocks new values. Thomas Alva Edison has probably the maximum number of 1093 patents to his credit. He quoted *"Anything that would not sell I do not want to invent. Its sale is proof of utility, and utility is success".*

Albert Einstein said "Intuitive mind is a sacred gift and rational mind is a faithful servant. We have created a society that honours the servant and has forgotten the gift". Thus IP created is product of intuitive mind. One of the objectives of IPRs is not to reinvent the things and waste the energy and at the same time creators of IPs are given the right to exploit their creations for a certain period of time

IPRs have been created to ensure protection against unfair trade practices. Owners of IPs are granted protection by a state and/or country under varying conditions and periods of time. This protection includes the right to: (i) defend the property they have created; (ii) prevent others from taking advantage of their ingenuity; (iii) encourage their continuing innovativeness and creativity; and (iv) assure the world a flow of useful, informative and intellectual works.

The convention establishing the World Intellectual Property Organization (WIPO) in 1967, one of the specialized agencies of the United Nations System, provided that "Intellectual Property" shall include rights relating to:

i. Literary, artistic and scientific works;

ii. Performance of performing artists, phonograms and broadcasts;

iii. Inventions in all fields of human endeavor;

iv. Scientific discoveries;

v. Industrial designs;

vi. Trademarks, service marks, and commercial names and designations; and

vii. Protection against unfair competition and all other rights resulting from intellectual activity in the industrial, scientific, literary, or artistic fields.

2.   Protection of Intellectual Property

Intellectual property is protected and governed by appropriate national legislations. The national legislations specifically describes what can be protected and under which legislation.

With the growing recognition of IPR, the importance of worldwide forums on IPs is realized. Companies, universities and industries want to protect their IPRs internationally. In order to reach this goal, countries have signed numerous agreements and treaties.

3.   Forms of Protection

Usually IPRs are protected by the following legal mechanism:

- Patents
- Copyrights
- Trademarks
- Trade secrets
- Geographical indications
- Designs
- Layout design of integrated circuits

Of these, patents are the most important forms of protection for research and development organizations. One of the most important examples of IP is the processes and products that result from the development of genetic engineering techniques. Another example of IP is the development of crop varieties, which are protected through Plant Breeder Rights or PBRs. Thus, a better understanding of patents as the form of IP by research scientists and university/institute administrators will increase the pace of research for technological developments in various fields of technology including biotechnology.

4.   World Organizations

4.1 GATT

GATT (General Agreement on Tariffs and Trade) was signed by 23 nations on October 30, 1947 at Geneva and took effect on January 1, 1948. GATT was meant to be a temporary arrangement to settle amicably among countries disputes regarding who gets what share of the world trade. GATT serves as a code of rules for international trade and a forum to discuss and find solutions to trade problems of the member countries. Agriculture was included for the first time in the 8th round of GATT negotiations in 1986 at Uruguay. The Uruguay round discussions were started in 1986, but successfully concluded on December 15, 1993. The negotiations were signed in the form of an accord on April 15, 1994 at Marrakesh, Morocco, by 123 countries that lead to the formation of World Trade Organization (WTO) on 1 January 1995. The Uruguay round covered

15 distinct areas and the Agreement on Agriculture (AoA) forms a part of the final act of the Uruguay negotiations. The long term objective of the agreement is to establish a fair and market oriented agricultural trading system and that a reform process should be initiated through the negotiation of commitments on support and protection and through the establishment of strengthened and more operationally effective GATT rules and disciplines. The original GATT text (GATT 1947) is still in effect under the WTO framework, subject to the modifications of GATT 1994.

4.2 WTO

The WTO was established on 1st January 1995 and is responsible for making and enforcing rules for trade between nations. WTO marks a major change in global trade rules. As an organization, it replaced the General Agreement on Tariffs and Trade (GATT), which had been in existence since 1948. GATT was enforced on 1st Jan., 1948 and was meant to be a temporary arrangement to settle amicably among countries disputes regarding who gets what share of the world trade.

WTO is the rule-based body for all trade and trade-related issues. It seeks to reduce barriers to trade through mutually advantageous agreements. It differs from GATT and other bodies in that it (i) covers all aspects associated with trade with binding rules for all the members, (ii) has a dispute settlement board, and (iii) has a built in agenda for review.

The role of WTO is much more extensive than that of GATT, which dealt with trade in goods. Apart from goods, the two other broad areas that WTO covers are services and intellectual property, which previously belonged to the domestic domain. Accordingly, WTO administers not only the Multilateral Trade Agreements (MTAs) in goods but also the General Agreement on Trade in Services (GATS) and the Agreement on Trade Related Aspects of Intellectual Property Rights (TRIPs), which came into existence with WTO. All the agreements annexed to the Agreement establishing the WTO were signed as part of a package deal. Member countries did not have the option of choosing some and rejecting others. Another important difference with the erstwhile GATT is that WTO has a stronger compliance mechanism than the GATT. A member's failure to meet the obligations can invoke retaliation across agreements and sectors.

WTO envisages a comprehensive scheme for protection of IPRs and the establishment of a legitimate reward system for the creative inputs of the inventors of intellectual property under the broad category of TRIPs. Various articles under TRIPs cover different forms of IP, *viz.* patents, trademarks, copyrights, designs, etc.

The long term objective of the agreement is to establish a fair and market oriented agricultural trading system and that a reform process should be initiated through the negotiation of commitments on support and protection and through the establishment of strengthened and more operationally effective GATT rules and disciplines.

Presently 162 states and customs territories ("WTO Members" or "Members") had joined the organization. India is a founder member of WTO. The WTO's substantive obligations are contained in a series of treaties relating to international trade, including 30 agreements on trade in goods, services, agriculture, sanitary and phytosanitary measures, trade-related investment measures, technical barriers to trade, textiles and clothing and intellectual property rights. These treaties are linked together as annexes to the agreement establishing the WTO. Through these agreements, WTO members operate a non-discriminatory trading system that spells out their rights and their obligations. Each country receives guarantees that its exports will be treated fairly and consistently in other countries' markets. Each promises to do the same for imports into its own market. The system also gives developing countries some flexibility in implementing their commitments. Disputes between WTO Members relating to any of these treaties are to be resolved by the WTO Dispute Settlement Body, which is composed of *ad- hoc* dispute settlement panels and a standing Appellate Body of trade experts. A WTO Member whose national laws or practices are challenged by another WTO Member and found to be incompatible with its WTO treaty obligations by the Dispute Settlement Body must modify those laws or practices or face the prospect of trade sanctions.

4.3 WTO-TRIPs

Agreement on Trade-Related Aspects of Intellectual Property Rights ("TRIPs" or the "TRIPs Agreement") was adopted in 1994 as a treaty administered by the WTO. As one of the WTO agreements, TRIPs is binding on all member countries of WTO. TRIPs aims at establishing strong minimum standards for intellectual property rights (IPRs). IPRs can be defined as the rights given to people over the creation of their minds. They usually give the creator an exclusive right over the use of his/her creation for a certain period of time. Intellectual property includes patents, copyrights, trademarks, geographical indications, industrial designs, integrated circuits and trade secrets. The protection of IPRs is binding and legally enforceable.

TRIPs is the first and only IPR treaty that seeks to establish universal, minimum standards of protection across the major fields of intellectual property. The TRIPs Agreement devotes minimal attention to plant breeders' rights or plant variety protection and does not even mention the UPOV Acts, but its adoption has encouraged the legal protection of plant varieties than any other international agreement.

4.4 World Intellectual Property Organization (WIPO)

The World Intellectual Property Organization ("WIPO") is one of the oldest specialized agency of the United Nations charged with promoting the protection of intellectual property throughout the world. The Paris Convention for the Protection of Industrial Property was born in 1883. This international agreement is the first major step taken to help creators ensure that their intellectual works are protected in other countries. The need for international protection of intellectual property (IP) became evident when foreign exhibitors refused to attend the International Exhibition of Inventions in Vienna, Austria in 1873 because they were afraid their ideas would be stolen and exploited commercially in other countries. The industrial property covers patents trademarks and designs. Following a campaign by French writer Victor Hugo and his *Association Littéraire et Artistique Internationale* the Berne Convention for the Protection of Literary and Artistic Works is agreed in 1886. The two secretariats set up to administer the Paris and Berne Conventions were combined to form WIPO's immediate predecessor, the United International Bureaux for the Protection of Intellectual Property known by its French acronym, BIRPI. The organization, with a staff of seven, is based in Berne, Switzerland. The Convention establishing the World Intellectual Property Organization (WIPO) came into force in 1970 and BIRPI was thus transformed to become WIPO. The newly established WIPO is a member state-led, intergovernmental organization, with its headquarters in Geneva, Switzerland.

The WIPO Secretariat undertakes wide variety of activities relating to IPRs, including hosting diplomatic conferences of government representatives seeking to negotiate new international treaties. WIPO's staff also provides technical assistance and training to member states and their national intellectual property offices, especially in developing countries. More recently, WIPO has created standing, expert and intergovernmental committees that conduct studies on particular intellectual property topics and generate non-binding guidelines and recommendations for consideration by WIPO members. WIPO administers 26 treaties including WIPO convention. Treaties administered by WIPO are given in Table 1. Details of the treaties are given in concerned chapters.

4.5 UPOV

In 1961 at Paris, a "Union International Pour La Protection Des Obtentions Vegetables" UPOV (International Union for the Protection of New Varieties of Plants) was signed for coordinating the inter country implementation of Plant Breeders Rights (PBR) and it entered into force in 1968 with its headquarters based at Geneva (Discussed in detail – Chapter 11).

Table 1.1: Treaties administered by WIPO

S.No	Treaties Administered by WIPO	Year
1.	Convention Establishing the World Intellectual Property Organization	1970
2.	Paris Convention for the Protection of Industrial Property	1883
3.	Berne Convention for the Protection of Literary and Artistic Works	1886
4.	Madrid Agreement for the Repression of False or Deceptive Indications of Source on Goods	1989
5.	Madrid Agreement Concerning the International Registration of Marks	1996
6.	Madrid Protocol Concerning the International Registration of marks	1996
7.	Hague Agreement Concerning the International Deposit of Industrial Designs	1925
8.	Nice Agreement Concerning the International Classification of Goods and Services for the Purposes of the Registration of Marks	1957
9.	Lisbon Agreement for the Protection of Appellations of Origin and their International Registration	1958
10.	International Convention for the Protection of Performers, Produccrz of Phonograms and Broadcasting Organizations (Rome Convention)	1961
11.	Locarno Agreement Establishing an International Classification for the Industrial Designs	1971
12.	Patent Cooperation Treaty	1978
13.	Strasbourg Agreement Concerning the International Patent Classification	1979
14.	Convention for the Protection of Producers of Phonograms Against Unauthorized Duplication of their Phonograms, Geneva (The Phonograms Convention)	1971
15.	Vienna Agreement Establishing an International Classification of the Figurative Elements of Marks	1973
16.	Brussels Convention Relating to the Distribution of Programme-Carrying Signals Transmitted by Satellite, Brussels	1974
17.	Budapest Treaty on the International Recognition of the Deposit of Microorganisms for the Purposes of Patent Procedure	1973
18.	Nairobi Treaty on the Protection of the Olympic symbol	1981
19.	Beijing Treaty on the International Registration of Audiovisual Works	2012
20.	Trademark Law Treaty	1994
21	Singapore Treaty on the Law of Trademarks	2006
22.	WIPO Copyright Treaty	1996
23.	WIPO Performance and Phonogram Treaty	1996
24.	Marrakesh Treaty to Facilitate Access to Published Works for Persons who are Blind, Visually Impaired or otherwise Print Disabled	2013
25.	Washington Treaty on Intellectual Property in Respect of Integrated Circuits	1989
26.	Patent Law Treaty	2000

Patents

Patent is a grant of an exclusive right for an invention by the Government to an inventor to prevent others from practicing i.e. making, using or selling the invention in exchange for full disclosure of the invention. The word patent is derived from the Latin word 'Patere" which means 'to open'. A patent is a personal property, which can be licensed or sold like any other property.

1. History of Patent Legislation

The first law providing these exclusive rights to an inventor dates back to 15th century in Italy. A Venetian statute was used to protect new and inventive devices that had been put into practice. In 1421, Filippo Brunelleschi in Florence became the first person to acquire an industrial patent. It gave him a three-year monopoly on the manufacture of a barge with hoisting gear which was used to carry marble from the Carrara quarries to the gates of Florence, sailing upstream on the Arno. The first U.S. Patent law was enacted in 1790. Since then, the U.S. Congress has enacted a variety of laws relating to patents, such as the 1930 Plant Patent Act. US patent law is grounded in Article I, Section 8, Clause 8 of the US constitution. Title 35 of the United States Code codified the Patent Act of 1952. Much of the biotechnology falls beneath the 'composition of matter' portion of §101.

There are three milestones in European patent legislation history – The Paris Convention in 1883, The Strasbourg Convention in 1963 and The Rectification of European Patent Convention in Munich in 1973 as part of the effort towards the establishment of a common market in Europe. Supervised by the Administrative Council the European Patent Office (EPO) is the administrative body of the EPC responsible for granting European patents.

1.1 Indian Patent System

The first legislation in India relating to patents was the Act VI of 1856. The objective of this legislation was to encourage inventions of new and useful manufactures and to induce inventors to disclose secret of their inventions. The Act was subsequently repealed by Act IX of 1857 since it had been enacted without the approval of the British Crown. Fresh legislation for granting 'exclusive privileges' was introduced in 1859 as Act XV of 1859. In 1872, the Act of 1859 was consolidated to provide protection relating to designs. It was renamed as "The Patterns and Designs Protection Act" under Act XIII of 1872. This Act was amended in the year 1883, with certain modifications in the patent law which were made in United Kingdom and it was considered that those modifications should also be incorporated in the Indian law. In 1888, an Act was introduced to consolidate and amend the law relating to invention and designs in conformity with the amendments made in the U.K. law.

The Indian Patents and Designs Act, 1911, (Act II of 1911) replaced all the previous Acts. This Act brought patent administration under the management of Controller of Patents for the first time. This Act was further amended in 1920, 1930 and 1945. After Independence, it was felt that the Indian Patents and Designs Act, 1911 was not fulfilling its objective. It was found desirable to enact comprehensive patent law owing to substantial changes in political and economic conditions in the country.

Accordingly, the Government of India constituted a committee under the Chairmanship of Justice (Dr.) Bakshi Tek Chand, a retired Judge of Lahore High Court, in 1949 to review the patent law in India. The committee submitted its interim report on 4th August, 1949 with recommendations for prevention of misuse or abuse of patent rights in India and suggested amendments to sections 22, 23 & 23A of the Patents & Designs Act, 1911 on the lines of the United Kingdom Acts 1919 and 1949. Based on the recommendations of the Committee especially for working of inventions and compulsory licence/ revocation, the 1911 Act was amended in 1950 and 1952 and finally a bill was introduced in the Parliament in 1953 (Bill No.59 of 1953). However, the Government did not press for the consideration of the bill and it was allowed to lapse.

In 1957, the Government of India appointed Justice N. Rajagopala Ayyangar Committee to examine the question of revision of the Patent Law and advise government accordingly. The report of the Committee, which comprised of two parts, was submitted in September, 1959. This report recommended major changes in the law which formed the basis of the introduction of the Patents Bill, 1965. This bill was introduced in the Lok Sabha on 21st September, 1965, which however lapsed. In 1967, again an amended bill was introduced which was referred to a Joint Parliamentary Committee and on the final recommendations of the Committee, the Patents Act, 1970 was passed. This Act repealed and replaced the 1911 Act so far as the patents law was concerned. However, the 1911 Act continued to be applicable to designs. Most of the provisions of the 1970 Act were brought into force on 20th April 1972 with publication of the Patent Rules, 1972.

This Act remained in force for about 24 years without any change till December 1994. An ordinance effecting certain changes in the Act was issued on 31st December 1994, which ceased to operate after six months. Subsequently, another ordinance was issued in 1999. This ordinance was subsequently replaced by the Patents (Amendment) Act, 1999 that was brought into force retrospectively from 1st January, 1995. The amended Act provided for filing of applications for product patents in the areas of drugs, pharmaceuticals and agrochemicals though such patents were not allowed. However, such applications were to be examined only after 31st Dec., 2004. Meanwhile, the applicants could be allowed Exclusive Marketing Rights (EMR) to sell or distribute these products in India, subject to fulfillment of certain conditions. The second amendment to the 1970 Act was made through the Patents (Amendment) Act, 2002 (Act 38 of 2002). This Act came into force on 20th May 2003 with the introduction of the new Patent Rules, 2003 by replacing the earlier Patents Rules, 1972. The third amendment to the Patents Act 1970 was introduced through the Patents (Amendment) Ordinance, 2004 w.e.f. 1st January, 2005. This Ordinance was later replaced by the Patents (Amendment) Act 2005 (Act 15 of 2005) on 4th April, 2005 which was brought into force from 1st January, 2005.

Under the provisions of section 159 of the Patents Act, 1970 the Central Government is empowered to make rules for implementing the Act and regulating patent administration. Accordingly, the Patents Rules, 1972 were notified and brought into force w.e.f. 20.4.1972. These Rules were amended from time to time till 20 May 2003 when new Patents Rules, 2003 were brought into force by replacing the 1972 rules. These rules were further amended by the Patents (Amendment) Rules, 2005 and the Patents (Amendment) Rules, 2006 which became effective from 5th May 2006. In 2014 Government notified Patents (Amendment) Rules, 2014. The important amendments are: a) The fee structure for filing of patent application as well as other proceedings before the Patent Office has been revised; b) A third category of Applicant for patent has been introduced in the form of "small entity" and the fees charged to them has been fixed; c) Every new application has to be accompanied by a newly introduced Form-28; d) Filling patent application and documents through the physical mode as opposed to the online mode will attract an additional fee of 10%. Recently some new draft guidelines of rules have been made: a) incentives are to be given to the inventors if India is made as international searching authority; b)filing of documents after First examination report period to be reduced to 6 + 3 months and c) start up industries are to be given incentive.

George Alfred De Penning is supposed to have made the first application for a patent in India in the year 1856. On February 28, 1856, the Government of India promulgated legislation to grant what was then termed as "exclusive privileges for the encouragement of inventions of new manufactures" i.e. the Patents Act. On March 3, 1856, a civil engineer, George Alfred De Penning of 7, Grant's Lane, Calcutta petitioned the Government of India for grant of exclusive privileges for his invention — "An Efficient Punkah Pulling Machine". On September 2, De Penning, submitted the specifications for his invention along with drawings to illustrate its working. These were accepted and the invention was granted the first ever Intellectual Property protection in India.

2.   International Treaties and Conventions

2.1   *Paris Convention*

The Paris Convention is an international convention for promoting trade among the member countries, devised to facilitate protection of industrial property simultaneously in the member countries without any loss in the priority date. All the member countries would provide national treatment to all the applications from other member countries for protection of industrial property rights. The Convention was first signed in 1883. Since then the Convention has been revised several times, in 1900 at Brussels, in 1911 at Washington, in 1925 at the Hague, in 1934 at London, in 1958 at Lisbon and in 1967 at Stockholm. The last amendment took place in 1979. India became a member of the Paris Convention on December 7, 1998. (Readers may note the use of the phrase 'Industrial Property' and not Intellectual Property). The principal features of the Paris Convention are:

1. National treatment
2. Right of priority
3. Independence of patents
4. Parallel importation
5. Protection against false indications and unfair competition

National treatment is a very important concept and is essential for successfully achieving the fundamental aims of the Paris Convention. The idea is to provide in a given member country, equal treatment to applications from member countries, and not to differentiate between the nationals of your country and nationals of the other countries for the purpose of grant, and protection of industrial property in your country. For example, a national of country X applies for grant of a patent in India. According to the Paris Convention, the Indian Patent Office shall apply the same norms and rules, to the applicant from X, as applicable to an Indian applicant, for granting a patent. Similarly the applicant from X shall have the same protection after the grant and identical legal remedies against any infringement shall be available to the applicant provided the conditions and formalities imposed upon Indians are complied with. No requirement as to domicile or establishment in the country where protection is claimed, may be imposed.

Right of priority means the date from which patent rights deemed to start is usually the date of filing of complete specification. To obtain rights in other member countries, the application must be filed on the same day in other member countries if it is desired to have the rights started from the same day. However, there are practical difficulties in synchronizing the activities. For facilitating simultaneous protection in member countries, the Paris Convention provides that within 12 months of national filing, if patent applications are filed in those member countries, the patents, if granted in member countries, will be effective from the date of national filing. This right is known as the right of priority. In other words you maintain the priority or the same date of filing in all the member countries and no one else in those countries can obtain the patent rights on a similar/identical invention from the same or a later date.

2.2 *Patent Cooperation Treaty and International Patenting*

PCT is a multilateral treaty entered into force in 1978. Through PCT, an inventor of member country (contracting state) of PCT can simultaneously seek patent protection for his/her invention in all/any of the member countries, without having to file a separate application in the countries of interest, by designating them in the international PCT application. India became a member of this treaty on Dec 7, 1998. PCT does not provide an institutional authority for granting international patents. There is, in fact, nothing like an international patent or world patent, which may cover all countries or at least all those who are members of the PCT.

The PCT facilitates obtaining patents in several countries relatively quickly. It standardizes the patent search and examination process so that the initial evaluation done in the country of filing is valid in all member countries. The principal objective of the PCT is to simplify the patenting system and to render more effective and more economical services by the offices which have responsibility for administering it in the interests of the users. To achieve its objective, the PCT:

i. Establishes an international system which enables the filing of applications to a single Patent Office in one language having similar effect in each of the countries which are party to the PCT and which the applicant names in his application;

ii. Subjects each international application to an international search which results in a report citing the relevant prior art which may have to be taken into account in deciding whether the invention is patentable; that report is made available first to the applicant and is later published;

ii. Provides for centralized international publication of international applications with the related international search reports, as well as their communication to the designated Patent Offices; and

iv. Provides the option of an international preliminary examination of the international applications to decide whether the claimed invention meets certain international criteria for patentability.

Further, PCT helps to facilitate and accelerate access by industries and other interested sectors to technical information related to inventions and to assist developing countries on gaining access to technology.

The patent office or any other office designated by each contracting state becomes a receiving office for receiving patent applications. The applications are referred to International Searching Authorities (ISA) which usually the patent offices, appointed to carry out the patent search on a global basis. In case the receiving office is also an ISA, a separate referral is not required. There is also a provision to get a patent application examined by International Preliminary Examining Authorities which, in most cases are ISA.

Under PCT, a patent application is first filed either in the home country or another country on the choice of the applicant. The National phase follows the International phase. Before the processing and examination begins in the designated office, the applicant has to enter the National phase otherwise International application loses its effect in the designated states. Any application can enter National phase within 31 months from the international filing date. In India Form I A has to be filled for National phase entry. It is not mandatory to submit the documents while entering the National phase since it is obligatory on the part of World Intellectual Property Organization (WIPO) to send the required document to designated office. The applicant has to deposit the national fee, usually the same as the fee required for the filing of a national or regional application.

In the normal case, if applicant is not taking the PCT route or a country is not a member of PCT then an applicant has to do multiple filing in different countries which makes the task difficult, besides expenditure on filing, search, translation, etc. In contrast, PCT establishes an international system which requires filing of only one additional application which is called "International Application" (IA) at a designated patent office available in all PCT member countries and patent can be written in one of the languages approved in the PCT country where the IA is filed.

The international search is a high quality search of the patent documents and other technical literature, at least in those languages in which most patent documents are filed (Chinese, English, French, German, Japanese, Russian and Spanish). The PCT prescribes high international standards for the documentation to be consulted from the qualified staff and effective search methods of the International Searching Authorities. There are six International Search Authorities: Australian Patent office (AU), Austrian Patent Office (AT), European Patent Office (EP), China Intellectual Property Office (CN), United State Patents and Trade mark Office (US) and Swedish Patent Office (SB).

The results of the search are set out in the 'international search report, which is made available by the 4th or 5th month after the international patent application is filed. It lists citations of prior art relevant to the claims of the application and gives an indication of the possible relevance of the citation to the questions of novelty and inventive step (non-obviousness). There is also an option of an international preliminary examination, to be made on the basis of the international search report, according to internationally accepted criteria of patentability (novelty, inventive step and utility).

It is to be noted that in case of applications for patent filed under PCT, it is mentioned as "International application" with the symbols as "WO", this does not mean that these applications relate to the protection of the invention disclosed in the application internationally. Accordingly there is nothing like 'World patent or PCT patent'. The patents granted under the PCT are individual national patents only and not world or international patents.

2.3 Strasbourg Agreement Concerning the International Patent Classification (1979)

The International Patent Classification, which is commonly referred to as the IPC, is based on an international multi-lateral treaty administered by WIPO. This treaty is called the Strasbourg Agreement Concerning the International Patent Classification, which was concluded in 1971 and amended in 1979. The Agreement is open to States party to the Paris Convention for the protection of Industrial Property. Although only 62 States are party to the Agreement, the IPC is used by the patent offices of more than 100 States, four regional offices and the Secretariat of WIPO in administering the Patent Cooperation Treaty (PCT) (1970). The Strasbourg Agreement establishes the International Patent Classification which, divides technology into eight sections with approximately 70,000 subdivisions. Each subdivision has a symbol consisting of Arabic numerals and letters of the Latin alphabet.

Example: **H01S 3/02**

H – Section [There are 8 sections A to H]

01 – Class symbol [Each class symbol consists of the section symbol followed by a two-digit number].

S – Sub class symbol [Each subclass symbol consists of the class symbol (01) followed by a capital letter].

3/02 – Each main group symbol consists of the subclass symbol followed by a one- to three-digit number, the oblique stroke and the number 02.

The appropriate IPC symbols are indicated on each patent document (published patent applications and granted patents), of which about 1,000,000 were issued each year in the last 10 years. The IPC symbols are allotted by the national or regional industrial property office that publishes the patent document. The Classification is indispensable for the retrieval of patent document in the search for "prior art." Such retrieval is needed by patent-issuing authorities, potential inventors, research and development units, and others concerned with the application or development of technology. In order to keep the IPC up to date, it is continuously revised and a new edition is published every five years.

2.4 Budapest Treaty

This is an international convention governing the recognition of microorganisms deposits in officially approved culture collections for the purpose of patent applications in any country that is a party to it. Because of the difficulties and on occasion of virtual impossibility of reproducing a microorganism from a description of it in a patent specification, it is essential to deposit a strain in a culture collection centre for testing and examination by others. The Treaty was signed in Budapest in 1973 and later on amended in 1980. India has become a member of this Treaty with effect from December 17, 2001.

3.   Patentable Subject Matter

Patent under Indian Patents Act, 1970 is granted for an invention which is defined as "a new product or process involving an inventive step and capable of industrial application while "inventive step" means a feature of an invention that involves technical advance as compared to the existing knowledge or having economic significance or both and that makes the invention not obvious to a person skilled in the art.

A patent is territorial in nature i.e. within the boundaries of a particular country, which has given the patent. The purpose of a patent is to encourage and develop new innovations. The Patent Law recognizes the exclusive right of a patentee to gain commercial advantage out of his invention. As the patent right is conferred by the State, it can be revoked by the State under very special circumstances even if the patent has been sold or licensed or manufactured or marketed in the meantime.

3.1   *Criteria for Patentable Invention*

There are three criteria to issue a patent for the innovation.

i. *Novelty*: Novelty is assessed in global context. An invention will be considered novel if it does not form a part of the global state of the art. The inventor must establish that the invention is new or novel. The novelty requirement refers to the prior existence of an invention. If an invention is identical to an already patented invention, the novelty requirement is not met, so a patent cannot be issued. An invention will not be novel if it has been disclosed in the public through any type of publications in the magazines, technical journals, books, newspapers, etc., or in radio, T.V. anywhere in the world before the filing of a patent application. Novelty is determined through extensive literature and patent database searches. It should be realized that patent search is the most crucial parameter for ascertaining the novelty. General principle is that it has not been anticipated by publication in any document anywhere in the world or form part of the knowledge, oral or otherwise, available within any local or indigenous community in India or elsewhere before the date of filing of patent application or date of priority.

ii. *Inventiveness (Non-obviousness)*: An invention is assessed for inventive step. The invention is not considered to involve an inventive step if it is obvious to a person skilled in the art on the date of priority in the light of the prior publication/knowledge/ document. It is an invention and not merely discovery. [An invention is a unique or novel device, method, composition or process. The invention process is a process within an overall engineering and product development process. It may be an improvement upon a machine or product, or a new process for creating an object or a result. While discovery is the process of finding information, a place, or an object, especially for the first time, e.g. a mineral present in the nature or a living being present in the nature which is already in existence]. The non-obvious requirement refers to the level of difficulty required to invent the technology. If an invention is so obvious that anyone having an ordinary skill would have thought of it, then it does not meet this requirement. *"Inventive step"* means a feature of an invention that involves technical advance as compared to the existing knowledge or having economic significance or both and that makes the invention not obvious to a person skilled in the art. Further, section 2(1)(l) defines "new invention" as any invention or technology which has not been anticipated by publication in any document or used in the country or elsewhere in the world before the date of filing of patent application with complete specification, i.e. the subject matter has not fallen in public domain or that it does not form part of the state of the art.

iii. *Usefulness (Industrial application)*: Invention should be capable of industrial application. It has a utility or is useful for the society. The useful requirement refers to the practical use of invention. If an invention provides a product that is required or needed in some manner, then it meets this requirement. "Capable of Industrial application", in relation to an invention, means that the invention is capable of being made or used in an industry.

In the patent adequate disclosure should be made so that others can also work on it. It should have the features: i) be a written description; ii) enables other persons to follow; iii) adequate and iv) deposit mechanism.

The purpose of a patent is to promote the progress of science and useful arts. The patent law promotes this progress by giving the inventor the right of exclusion. In exchange for this right to exclude others, the inventor must disclose all details describing the invention, so that when the patent period expires, the public may have the opportunity to develop and profit from the use of invention.

There are three types of patents in USA: (i) plant patent, (ii) design patent, and (iii) utility or regular patent.

Plant patents are granted for newly developed asexually propagated plants under Plant Patent Act 1930 in USA. It provides protection for 20 years. A design patent protects ornamental characteristics. The life span of a design patent is only 14 years. Examples of design patents are the companies dealing with toys, souvenirs and industrial manufacturers. Utility patent is the most common patent used by universities and companies to protect their results of research and development. In USA and most of the countries it has a life span of 20 years from the date of filing an application as it is one of the requirements of TRIPs regulations.

Most biotechnology inventions are filed as utility patents and not as plant patents. As a utility patent, it is possible to protect plant genes and other controlling elements rather than just the plant, and to control the use of genetic material of a number of plants and for multiple uses such as disease resistance, herbicide resistance, or pharmaceutical or oil production.

3.2 Non-patentable Inventions

An invention although may satisfy the conditions of novelty, inventiveness and usefulness but there are some inventions which have been categorized under non-patentable inventions in India (Sections 3 and 4 of the Patents Act, 1970 amendment 2005). The following are the non-patentable inventions under Sec. 3 of the Act:

a. An invention which is frivolous or which claims anything obvious contrary to well established natural laws.

b. An invention the primary or intended use or commercial exploitation of which could be contrary to public order or morality or which causes serious prejudice to human, animal or plant life or health or to the environment.

[Examples: Any device, apparatus or machine or method for committing theft/burglary, counterfeiting of currency notes, method for gambling; a device for house-breaking; a method of adulteration of food, Terminator gene technology, etc.]

c. The mere discovery of a scientific principle or the formulation of an abstract theory or discovery of any living things or non living substances occurring in nature.

[Examples: A discovery is non patentable subject matter e.g. Vasco de Gama discovered India, Columbus discovered America are non patentable and likewise finding of a new substance or micro-organism or organism occurring freely in nature is a discovery. It adds to the amount of human knowledge by disclosing something already existent, which has not been seen before. Whereas an invention adds to the human knowledge by creating a new product or process involving a technical advance as compared to the existing knowledge].

d. The mere discovery of a new form of a known substance which does not result in the enhancement of the known efficacy of that substance or the mere discovery of any new property or new use for a known substance or of the mere use of a known process, machine or apparatus unless such known process results in a new product or employs at least one new reactant. (For the purpose of this clause, salts, esters, ethers, polymorphs, metabolites, pure form, particle size, isomers mixtures of isomers, complexes, combinations and other derivatives of known substance shall be considered to be the same substance, unless they differ significantly in properties with regard to efficacy).

This Section 3(d) is one of the most important sections and challenged in the court of law and many representations have been made by MNCs to delete this section [See Boxes 1 and 2].

The different examples under this section have been given below:

i. Paracetamol drug has antipyretic property. Further discovery of new property of paracetamol as analgesic cannot be patented. Similarly, ethyl alcohol is used as solvent but further discovery of its new property as anti knocking, thereby making it usable as fuel, cannot be considered patentable.

ii. A known substance in its new form such as amorphous to crystalline or crystalline to amorphous or hygroscopic to dried, one isomer to other isomer, metabolite, complex, combination of plurality of forms, salts, hydrates, polymorphs, esters, ethers, or in new particle size, shall be considered same as of known substances unless such new forms significantly differ in the properties with regard to efficacy. The requirement here is that the new form must result in enhancement of known efficacy of known substance and that in order to be distinct from the known substance, the new form must differ in the properties with regard to efficacy.

iii. In a claim relating to food packing machine, it was clear that it can pack food as well as any other compound also.

iv. Aspirin was used for analgesic purpose, but one cannot get a patent for its new use for treatment of the cardiovascular disease.

Box 1

Novartis – Glivec Anticancer Drug Case

India ratified WTO-TRIPs agreement on 1st Jan., 1995; therefore it was obliged to introduce a mailbox transition system in compliance to Article 70.8 of TRIPs. One application as a convention "mailbox application" was filed in India on 17th July, 1998 (at Patent Office, Chennai) by M/s Novartis AG, Switzerland with Indian Patent application No. 1602/MAS/1998 for the invention titled "Crystal modification of a N-Phenyl-2-Pyrimidineamine derivative, processes for its manufacture and its use". It sought patent protection for beta crystalline salt form of the free base, Imatinib, which was covered by an earlier pre-1995 patent No. 5521184A,10 popularly known as "Zimmerman Patent". Patent application was kept in the mailbox and not opened for examination until 2005. In the mean time on 10th Nov., 2003 Novartis obtained Exclusive Marketing Rights (EMR) for marketing Glivec in India. It started enforcing its rights and restrained generic manufacturers from manufacturing and selling the drug which resulted in increase of price from Rs 90 per 100 mg capsule to Rs 1000 per 100 mg capsule. After amendment to the Patents Act in 2005, the Government has introduced an explanation to sec 3(d) introducing the concept of 'efficacy'. Upon publication of the patent, the grant of patent was opposed by way of representation u/s 25(1) by M/s Natco Pharma Ltd., India on 26th May, 2005 and they also requested for hearing. The grounds for opposition were i) Anticipation by prior publication ii) Lack of inventive step iii) Non-patentability u/s 3(d) of the Patents Act and iv) Wrongfully claiming the priority. Applicant filed the reply statement with evidence on 25th July, 2005 and also asked for hearing. The application filed in India has claimed the Swiss priority dated 18th July, 1997, but Switzerland was not a Convention country on that date. It became the Convention country in September, 1998 only. Hence, no priority of Swiss application could be claimed in respect of the Indian application. The title compound commercially, called imatinib mesylate, was already known in the US Patent No. 5521184 (1993 Patent). The 1993 Patent disclosed methane sulphonic acid as one of the salt –forming groups and also states that the required acid additions salts were obtained in a customary manner. Another Document, "Nature Medicine" (5th May, 1996) also described the title compound. Also the compound, imatinib mesylate salt, inherently existed in the β-crystalline form, which is most stable form of the salt. Hence, the claims of the application for the product and process in respect of the title compound stood anticipated by prior publications. Since the 1993 patent disclosed the free base of the base compound, it was obvious to a person skilled in the art to prepare the corresponding pharmaceutically acceptable salts. Hence, the product claims were obvious. As per section 3(d), any salt or polymorph or derivative of the known substance is not patentable unless such salt or polymorph or derivative shows enhanced efficacy of the substance. As regards efficacy, the patent specification itself states that, wherever β-crystals are used, the imatinib free base or other salts can be used. The affidavit submitted by the technical expert on behalf of the applicant demonstrated that the relative bioavailability of the free salt with that of β-crystal form of imatinib mesylate differ only by 30% and accounted this difference to their solubility in water. Thus, the specification did not bring out any improvement in the efficacy of the β-crystal form over the known substances; rather it stated that the base compound could be used equally in the treatment of diseases or in the preparation of pharmacological agents wherever the β-form is used. Thus, the product claim amounted to a mere discovery of the new form of the known substance. Imatinib mesylate is a compound for anti cancer drug Glivec. Pursuant to the pregrant oppositions, on 25th Jan., 2006, the Patent Controller in Chennai refused to grant a patent to Novartis. Accordingly, the EMR got extinguished. Novartis case started in 2006-2007 and wended through India's Patent Controller, High Court, IPAB and Apex Court. On 1st April 2013- Supreme Court rejected Novartis plea which challenged Indian Patents Act, 1970 as non compliant to TRIPs. The Hon'ble Supreme Court held – 'that the patent product, the beta crystalline form of imatinib mesylate, fails in both the tests of invention and patentability as provided under clauses (j), (ja) of section 2(1) and section 3(d) respectively, the appeals filed by Novartis AG are dismissed with cost'. The multinational firm in question just hasn't mustered up convincing evidence to refute that it is guilty of 'evergreening' Glivec with incremental chemical changes that are without similarly incremental therapeutic value to keep the pill costly and out of reach for most people. [**Evergreening** refers to a variety of legal and business strategies by which technology producers with *patents* over products that are about to expire retain royalties from them, by either taking out new *patents* (for example over associated delivery systems, or new pharmaceutical mixtures), or by buying out or frustrating competitors, for longer periods of time than would normally be permissible under the law].

Box 2

Boehringer Denied Patent on HIV Drug – Nevirapine

The Indian Patent Office has denied Germany's Boehringer Ingelheim a patent on its key HIV drug, Nevirapine, for a version sold as Viramune XR (extended release), thwarting attempts by Big Pharma for "exclusivity" extension on their patented drugs to reportedly block entry of affordable generics. The application on the Nevirapine extended release formulation was refused since it was found to be "obvious" and lacking inventive step, violating Section 3(d) — an important safeguard in the patent law specifically relevant for pharma and chemical industries — prohibits grant of patents to new forms of known substances, unless the new form results in enhanced efficacy over the known substance. The MNC had applied for a patent in India on the extended release of the HIV drug in July 2009, on which domestic company Cipla filed a pre-grant opposition in 2011, and later launched its generic version. The application was examined by the Indian Patent Office, which raised objections in its examination report in January 2014 over obviousness and Section 3(d).

e. A substance obtained by a mere admixture resulting only in the aggregation of the properties of the components there of or a process for producing such substance.

[Examples: A mixture of different types of medicament or medicines to cure multiple diseases, a mixture of different pesticides or insecticides to control diseases or pests are mere admixture of known substances and are not patentable inventions. It is a patentable subject matter if this admixture acts in a synergistic manner and there is enhancement of its properties. Example: a composition of two drugs, i.e. Paracetamol and Ibuprofen for curing fever and pain or process of preparation thereof is not patentable for the reason that the composition is a mere admixture of two drug components resulting into aggregation of properties thereof. Since paracetamol is well known for treatment of fever and Ibuprofen for treatment of pain. However, if the mixture of drugs exhibits some unexpected results or synergistic properties in their action, then such composition is considered as patentable subject matter.]

f. The mere arrangement or re-arrangement or duplication of known devices each functioning independently of one another in a known way.

[Examples: An umbrella with fan, bucket fitted with torch, clock and transistor in a single cabinet. These are not patentable subject matter, since they are nothing but mere arrangement and rearrangement of items without having any working interrelationship between them and functioning independently of each other.]

g. *This section has been omitted in the new amendment Act 2005.*

h. A method of agriculture of horticulture.

[Examples: A method of growing leguminous plants as inter-cropping for improving fertility of soil by augmenting nitrogen content of the soil.

A method of producing a new form of a known plant, even if it involved a modification of the conditions; a method of producing mushroom plant or a method for cultivation of an algae were held not patentable]

i. Any process for the medicinal, surgical, curative, prophylactic, diagnostic, therapeutic or other treatment of animals to render them free of disease or to increase their economic value or that of their products.

[Examples: A method of treatment of malignant tumour cells and method of removal of dental plaque and caries are not patentable, since they are held as treatment of human beings. Treatment of sheep for increasing wool yield; Surgical methods: a stitch-free incision for cataract removal, a method of treatment of human or animal body by surgery or therapy or of diagnosis practiced on the human or animal body. Curative methods: a method of cleaning plaque from teeth. Any medical treatment of a disease, ailment, injury or disability, i.e., anything that is wrong with a patient and for which he/she would consult a doctor; Prophylactic treatments: vaccination and inoculation, are to be regarded as therapy. The same considerations apply for animals as well e.g. prophylaxis and immunotherapy in animals are regarded as therapy. An application of substance to human body purely for cosmetic purposes is not a treatment or therapy. On the other hand, application of an ointment to the skin designed to be effective to remove keratoges from the skin would be the instance of medical treatment.

Methods of treatment of the human or animal body by surgery are excluded. 'Surgery' is defined as the treatment of disease or injury by operation or manipulation. It is not limited to cutting the body but includes manipulation such as the setting of broken bones or relocating dislocated joints (sometimes called "closed surgery"), and also dental surgery. In general, any operation on the body, which required the skill and knowledge of a surgeon, would be regarded as surgery and includes non-curative treatments such as cosmetic treatment, the termination of pregnancy, castration, sterilization, artificial insemination, embryo transplants, treatments for experimental and research purposes and the removal of organs, skin or bone marrow from a living donor are, if carried out by surgery, regarded as surgical treatments. Once it has been decided that a method constitutes surgery, therapy or diagnosis practiced on the human or animal body, it is necessarily non-patentable. For example, methods of abortion, induction of labour, control of oestrus or menstrual regulation are always therapy, irrespective of the reason for the treatment.]

j. Plants and animals in whole or any part thereof other than microorganisms but including seeds, varieties and species and essentially biological processes for production or propagation of plants and animals. [Example: Claim: A method of producing at least one of substantially pure hybrid seeds, plants and crops, comprising the steps of (i) producing a male parent which is male fertile, (ii) breeding the male parent with a female parent which is substantially male sterile, and (iii) harvesting seeds from the female parent which contain pure hybrid seeds. Analysis: The claimed method involves the step of cross breeding for producing pure hybrid seeds, plants and crops. Thus, it is an essentially biological process and is not allowable.

k. A mathematical or business method or a computer program *per se* or algorithms.

l. A literary, dramatic, musical or artistic work or any other aesthetic creation whatsoever including cinematographic works and television productions.

   [Example: Writings, music, works of fine arts, paintings, sculptures, computer programmes, electronic databases, books, pamphlets, lectures, addresses, sermons, dramatic-musical works, choreographic works, cinematographic works, drawing, architecture, engraving, lithography, photographic works, applied art, illustrations, maps, plans, sketches, three-dimensional works relating to geography, topography, translations, adaptations, arrangements of music, multimedia productions, etc. are not patentable. Such works fall within the domain of the Copyright Act, 1957.]

m. A mere scheme or rule or method of performing mental act or method of playing game.

   [Example: Method of – learning a language, playing chess, teaching, learning, operating a machine or equipment as per set of instructions, etc.]

n. A presentation of information.

   [Example: Any manner, means or method of expressing information whether visual, audible or tangible by words, codes, signals, symbols, diagrams or any other mode of representation is not patentable. For example, a speech instruction means in the form of printed text where horizontal underlining indicated stress and vertical separating lines divided the works into rhythmic groups is held not patentable.]

o. Topography of integrated circuits.

   [Since protection of Layout Designs of Integrated Circuits is governed separately under the Semiconductor Integrated Circuit Lay-out Designs Act, 2000, three-dimensional configuration of the electronic circuits used in microchips and semiconductor chips is not patentable.]

p. An invention which in effect, is traditional knowledge (TK) or which is an aggregation or duplication of known properties of traditionally known component or components.

   [Examples: Anti-septic property of turmeric for wound healing. Pesticidal and insecticidal properties of *neem*. For the examination of TK related subject matters, separate guidelines have already been issued by the Office of Controller General of Patents, Designs and Trade Mark (CGPDTM)].

Section 4 states that "No patent shall be granted in respect of an invention relating to atomic energy falling within subsection (1) of section 20 of the Atomic Energy Act, 1962". The Central Government has notified the substances, equipment and technology which comes under the Act. The "prescribed substances" means any substance including any mineral which in its opinion is or may be used for the production or use of atomic energy or research e.g. uranium, plutonium, thorium, beryllium, neptunium, deuterium, heavy water (deuterium oxide) or any of these respective derivative or compounds or any other materials containing any of the aforesaid substances. The Act also defines the term "radioactive substances" or "radioactive material" as any substance or material, which spontaneously emits, radiation in excess of the levels prescribed by notification by the Central Government.

4.   Types of Patent Applications

There are different types of patent applications for grant of patent:

1. Ordinary application: If an application is made in the Patent Office without claiming any priority of application made in a convention country or without any reference to other application under process in the office is called an ordinary application. This is the normal routine application.

2. Convention application: When an application comes to the patent office claiming a priority date based on a similar application filed in one of the convention country, it is called a convention application. [A convention country is a member of a group of countries or a union of countries or an Inter-governmental organization which are notified by the Ministry of Commerce and Industry, Government of India under sub-section(1) of Section 133. As per the notification published in the Official Gazette on 20th May, 2003 there are 180 members].To get a convention status an applicant should file the application in Indian Patent Office within twelve months from the date of first filing of a similar application in one of the convention country(ies). A convention application should be accompanied by a complete specification.

3. PCT international application: The Patent Cooperation Treaty (PCT) is an international patent law treaty, concluded in 1970. It provides a unified procedure for filing patent applications to protect inventions in each of its contracting states. A patent application filed under the PCT is called an international application, or PCT application. A single filing of a PCT application is made with a Receiving Office (RO) in one language. It then results in a search performed by an International Searching Authority (ISA), accompanied by a written opinion regarding the patentability of the invention, which is the subject of the application. It is optionally

followed by a preliminary examination, performed by an International Preliminary Examining Authority (IPEA). Finally, the relevant national or regional authorities administer matters related to the examination of application (if provided by national law) and issuance of patent. A PCT application does not itself result in the grant of a patent, since there is no such thing as an "international patent", and the grant of patent is a prerogative of each national or regional authority. In other words, a PCT application, which establishes a filing date in all contracting states, must be followed up with the step of entering into national or regional phases to proceed towards grant of one or more patents. The PCT procedure essentially leads to a standard national or regional patent application, which may be granted or rejected according to applicable law, in each jurisdiction in which a patent is desired.

4. PCT national phase application: The national phase is the second of the two main phases of the PCT procedure. It follows the international phase and consists of processing of the international application before each Office of a Contracting State that has been designated in the international application. The International Bureau of WIPO publishes the application and the search report is available 18 months after the priority date. The original application is then sent to the designated offices indicated in the application. Within two months of the application i.e. by the 20th month, the applicant will have to formally apply to the national patent offices of those countries for grant of patent by paying official fees and completing other formalities. The national phase of processing the international application by the designated Office is generally delayed until the termination of the international phase on the expiration of the time limit which is 30 months. In other words , the 20th month period has been virtually extended to 30th or 31st months. Within two months of this, the applicant/inventor will have to formally apply for grant of patent in designated country. This is referred as PCT national phase application. The priority date is the same as the date of filing the original PCT application.

5. Application for patent of addition: When an applicant feels that he has an invention, which is a slight modification of the invention for which he has already applied in India, the applicant can go for a patent of addition rather than filing a new or separate application for that invention. In such cases applicant need not pay any separate renewal fee during the term of main patent. Patent of addition expires along with the main patent unless it is made independent.

6. Divisional application: When the application made by an applicant claims more than one invention, the applicant or the Controller may divide the application and file two or more applications for each of the invention. This type of application divided out of the one parent application is called divisional application. The priority date for all divisional application(s) will be the same as that for the parent application.

5.   Patent Grant Procedure

Filing of an application for a patent should be completed at the earliest possible date and should not be delayed unless the invention is fully developed for commercial working. Further, a published or disclosed invention cannot be patented. Hence inventors should not disclose their inventions in symposia or as research papers before filing a patent application. The preparation of a patent application is quite complex and generally an attorney is required to draft the application.

A patent is enforced in the country that issues it. For each country a separate application is to be filed in that country where protection is sought if a country is not a member of patent cooperation treaty (PCT). Application is required to be filed according to the territorial limits where the applicant or the first mentioned applicant in case of joint applicants, for a patent normally resides or has domicile or has a place of business or the place from where the invention actually originated. If the applicant for the patent or party in a proceeding having no business place or domicile in India, the appropriate office will be according to the address for service in India given by the applicant or party in a proceeding . The appropriate office once decided in respect of any proceedings under the Act shall not ordinarily be changed. In India Head Patent Office is located at Kolkata whereas branch offices for different zones are in Delhi, Mumbai and Chennai. Application may be made by the inventor, either alone or jointly with another inventor, or assignee of the inventor or legal representative of the inventor. A patent application is to be submitted in the prescribed form for one invention only. Every application shall accompany provisional or a complete specification. The document that contains the detailed description of invention alongwith the drawings and claims is called patent application with complete specifications. Claims refer to the scope of protection sought. From 20th July, 2007 the Indian Patent Office has put in place an online filing system for patent application. Before filing a patent application, an applicant must perform patent search for novelty of its invention.

Patent with complete specification is a document having the following components:

i. Title of invention

ii. Field of invention

iii. Background of invention with regard to drawbacks associated with prior art

iv. Object of invention

v. A summary of invention

vi. A brief description of the accompanying drawing if any

vii. Detailed description of the invention with reference to examples and drawing

viii. Claim(s)

5.1 Patent Search and Patent Databases

Before filing an application, patent search should be made to determine the novelty of an invention. A patent search is a library-type investigation made prior to the filing of a patent application. Its purpose is to find disclosures of the proposed invention, or similar inventions, in publications of technology or previous patent documents in order that a patent be granted. Doing a search before applying for a patent saves time in fruitless research and avoids the possibility of future infringement suits. Patent search can be done at the Patent Office, on the Internet, through private information databases. Some patent search resources are given in Table 2.1.

WIPO recently launched, 'WIPO GOLD', a free, on-line global intellectual property (IP) reference resource that provides quick and easy access to a broad collection of searchable IP data and tools relating to, for example, technology, brands, designs, statistics, WIPO standards, IP classification systems and IP laws and treaties. Powerful databases, such as WIPO's PATENTSCOPE® search service, make it possible to conduct, free-of-charge, high-quality searches of data relating to over 1.7 million international patent applications filed under the PCT, and patent data collections of a growing number of countries.

Table 2.1: Patent search resources

Resource	Internet address
USPTO – Patent and trademark office	www.uspto.gov/web/menu/search.html
EPO – European Patent Office	www.ep.espacenet.com www.european-patent-office.org
Delphion Intellectual Property Network	www.delphion.com/
The French Office (INPI) patent database	www.inpi.fr/inpi/inbrevet.html
Free Patent Search Site	www.alliance-DC.org/
Biotechnology patents	www.nal.usda.gov/bic/Biotech patents/
Indian patent office database	www.ipindia.nic.in www.indianpatents.org.in
Canadian patents	www.library.ubc.ca/patscan
WIPO patent site	www.wipo.org
US & EPO patents	www.patents.ibm.com
PCT applications since 1998	pctgazette.wipo.int
IPR Law Link India	www.iprlawindia.org
Australia's IP database	www.ipaustralia.gov.au
Database search services	www.patent.gov.uk

5.2   Patent Specifications

Basically there are two types of patent documents: Application with provisional specification or complete specification. A **Provisional Specification** application should contain the description of the invention with drawing(s) if required. But there are no claim(s). This document is submitted when inventor(s) is still working on the invention. This is being done to claim the priority of an invention. (For example inventor has filed patent application with provisional specification on 15th Jan., 2015). However, the complete specification including claims fairly based on the matter disclosed in the provisional specification should be filed within 12 months from the date of filing of provisional specification (e.g. filed on 8th Dec., 2015). The date of patent will be the date of filing (Priority date) of provisional specification (15th Jan., 2015). An application with **Complete Specification** should fully describe the invention with drawing(s) if required disclosing the best method known to the applicant and it ends with claim(s) defining the scope of protection sought. Writing of claim(s) is the most important component of the patent application. Claims are those portions of the patent that describe what can be accomplished with the invention and what is protected in the patent. This is neither to be too general nor to be too narrowly defined.

5.3   Patent Claims

It is the claims that define the boundaries of the patent owner's rights. The description in the patent specification needs to clearly state the claims. It determine if someone is infringing a patent, that is making, using, etc., without the patent owner's permission. The allegedly infringing product is compared only to the claims but never to what is mentioned in the title, abstract or in the general description. The purpose of the claims is to highlight the precision, conciseness and accuracy that how much of the invention is protected. The claim(s) of a complete specification shall relate to a single inventive step. The claim(s) have to be unitary, irrespective of the total number of such claims in a patent. If more than one claim is made, then each one has to be serially numbered and set in separate paragraphs, but without breaking the sentence/punctuating the full stop.

The first claim in any patent is called the 'principal claim' and the rest are 'subordinate claims'. The principal claim should comprehensively cover the scope and range of invention, defining its novel as well as other aspects. Highest care has to be taken for writing the patent claim as any aspect left out from the claim would not qualify the patent for legal protection on that aspect. Subordinate claims are made to describe certain aspects of principal/another subordinate claim. In writing such claims, a reference is always made to the principal or the relevant subordinate claim. Subordinate claim does not stand alone.

Some of the statements are not considered as the claim(s) like: I claim to be the inventor; I claim a patent and no one else shall use my invention; I claim that the machine/process/product described above is quite new and has never been used or seen before, etc.

6. Publication and Examination of Application

6.1 Publication

All the applications for patent, except the applications prejudicial to the defence of India or abandoned due to non-filing of complete specification within 12 months after filing the provisional or withdrawn within 15 months of filing the application, are published in the Patent Office Journal just after 18 months from the date of filing of the application or the date of priority whichever is earlier. The priority date of patent is the date of filing the document with complete specification. This is an important date because it is from this date that the legal protection of an invention covered in the patent takes effect. The term of the patent is counted from this date. The publication includes the particulars of the date of the application, application number, name and address of the applicant along with the abstract. The applications for patent are not open for public inspection before publication.

6.2 Request for Examination

An application for patent will be examined only if request is made by the applicant or by any other third person interested in the patent with prescribed fee. 10% fee is extra when it is physically filed. Where no request for examination of the application for patent has been filed within the prescribed period, the application will be treated as withdrawn and, thereafter, application cannot be revived.

6.3 Examination

Application will be taken up for examination, according to the serial number of the requests received on Form 18. A First Examination Report (FER) stating the objections/requirements is communicated to the applicant or his agent according to the address for service ordinarily within six months from the date of request for examination or date of publication whichever is later. Application or complete specification should be amended in order to meet the objections/requirements within a period of 12 months from the date of FER. If all the objections are not complied with within the period of 12 months, the application shall be deemed to have been abandoned. When all the requirements are met the patent is granted, after 6 months from the date of publication, the patent letter is issued, entry is made in the register of patents and it is notified in the Patent Office Journal.

The patent application can be withdrawn at least three months before the first publication which will be 18 months from the date of filing or date of priority whichever is earlier. The application can also be withdrawn at any time before the grant of the patent. However, the application withdrawn after the date of publication cannot be filed again as the invention is already known to the public and has lost the novelty. But application withdrawn before the publication can be filed again provided it is not opened to public otherwise.

If any patent application is to be filed abroad, either it should be filed first in India or without filing in India, it should be made only after taking a written permission from the Controller. The request for permission should be accompanied with a gist of invention along with the Form-25.

The CGPDTM has *vide* a Public Notice dated July 2, 2014 informed that deposition of biological material to an international depository authority by applicants of patents should be made prior to the date of filing of patent application in India and reference of such deposition in the patent application should be made within a period of three months from the date of filing of such application, in case the same is not already made as per Rule 13(8) of the Patents Rules, 2003. The concerned applications are liable to be refused under the Act for non-compliance of the aforesaid provisions.

It is a common experience that through ignorance of patent law, inventors act indiscreetly and jeopardize the chance of obtaining patents for their inventions. The most common of these indiscretions are: a) to publish their inventions in newspapers or scientific and technical journals, before applying for patent; b) to wait until their inventions are fully developed for commercial working, before applying for patents. Delay in making application for a patent involves risks, namely, i) other inventors might forestall the first inventor in applying for the patent, and ii) there might be either an inadvertent publication of the invention by the inventor himself, or the publication thereof by others independently of him.

7. Term of Patent

The term of a patent in India is 20 years from the date of filing of complete specification of patent application, as per The Patents Act, 1970 amendment 2005. In spite of date of filing being the date of patent, a suit or proceeding cannot be commenced or prosecuted against infringement committed before the date of publication. The term of patent in case of International applications filed under the PCT designating India, will be 20 years from the International filing date accorded under the PCT.

A patent is maintained by paying the maintenance fee every year. If the maintenance fee is not paid, the patent will cease to remain in force and the invention becomes open to public. Anyone can then utilize the patent without the danger of infringing the patent. However, if a patent lapses on account of non- payment of maintenance fee, it can be restored upon making an application in the prescribed manner to The Controller of Patents.

8. E-filing of Patent Applications

CGPDTM has launched E-filing Version 1.0 of patent applications from July 2007 which was updated to E-filing Version 2.0 (comprehensive filing) from Dec., 2012. Comprehensive version has all the features of authenticity, identity and security. E-filing is a service provided by the Intellectual Property Office,

India in order to enable customers to apply for a patent online allowing from the User's browser for the User to complete an electronic application form with the associated attachments and necessary payment details.

8.1 Procedure for e-filing

1. Acquire Class 3 Digital Signature(s) either from (n) Code Solutions, Tata Consultancy Services (TCS) & Safe Script.

2. New users (Applicants, Agents or Attorneys), can complete online registration by providing Digital Signature details to get a User ID and Password for using the e-Filing System of Indian Patent Office (IPO).

3. Secure Login into the system with created User Id and the Password.

4. Download the client software for preparing patent application offline with required documents and digitally sign it for uploading on IPO Server.

5. Fill patent application offline and generate an XML file using Client Software.

6. After creating application (XML) file offline, digitally sign the XML file (Max. file size permitted 5MB) for uploading on to the IPO Server.

7. Login into e-Patent portal (http://ipindia.gov.in) for uploading application XML file on IPO Server.

8. Upload & submit digitally signed XML file to IPO server.

9. Process Application for EFT (Electronic Fund Transfer).

10. Review application status on e-Patent Portal.

11. On successful EFT acknowledgement details would be displayed/generated.

12. Print acknowledgement. Click on "Print" to generate printout of acknowledgement.

The CGPDTM has launched a new online feature **Dynamic Utilities** on real time basis which allow stakeholders to:

- Conduct searches on granted and published applications in Patent/ Trademark databases
- Conduct searches to ascertain the status of patent/trademarks
- Access the desired information on real time basis
- Search information using combination of various search parameters
- Access on real time basis to examination reports issued by the office.

Thus, Dynamic Utilities help in increasing transparency and dissemination of IP Information. It helps to view request for examination for which FER is issued, disposal of patent applications granted, refused and abandoned during the specific period, patents working, patents expired and pending and grant status of patents. This feature was earlier available only for trademarks but is now available for patents as well. The said facility can be accessed at: http://ipindiaservices.gov.in/ePatentFiling/.

9. Use of Technical Information in Patent Documents

Patent literature has an important use as a source of useful scientific information:

A) It is useful in searching for new technical solutions in the course of research and development and the creation and assimilation of new technology.

B) A scientist or technologist when beginning a new project should first study the patented literature and other information sources to find out what should not be invented, what technical solutions have already been found, when, by whom and where. The timely use of patent information makes it possible to avoid unnecessary expenditure and save time and resources. It also provides information on the trends of the development of knowledge in that field of research.

10. Opposition and Revocation of Patent

Opposition: There can be both pre and post grant opposition. *Pre grant opposition* can be made by any person to the Controller at an appropriate office for a patent application which has been published but not granted within three months from the date of such publication (section 25) or before the grant of patent, whichever is later. The opposition may be filed on the grounds of patentability including novelty, inventive step and industrial applicability by non-disclosure or wrongful mentioning in complete specification, source and geographical origin of biological material used in the invention and anticipation of invention available within any local or indigenous community in India or elsewhere. *Post grant opposition* can be made if a patent is a granted one but the period of one year from the date of publication of grant of a patent is not over then any person interested may give notice of opposition to the Controller [section 25(2)]. Post grant opposition can be filed on any of the following grounds:

1. Wrongfully obtained the invention.
2. Prior publication of the claimed invention before the priority date of the claims.
3. If the claimed invention in the complete specification is publicly known or used in India before priority date.
4. If the claims does not involve any inventive step.
5. If the invention is non patentable under the Act.
6. If the complete specification does not describe the invention or the method by which it is to be performed.

7. In case of convention application, the application for patent has not been made within twelve months from the date of first application used for protection.

8. If the complete specification does not disclose or wrongfully mention the source and geographical origin of the biological material used for invention.

In case of post grant opposition, Controller constitutes a Board called Opposition Board. Patentee and the opponent are given an opportunely of being heard and on the basis of recommendations of Opposition Board, the Controller shall order either to maintain or to amend or to revoke the patent.

Revocation: A patent granted does not mean that all the claims made in the specifications are valid. The validity of a patent can be challenged even after it is granted. Such kind of objections raised after the grant of patent is called revocation. A revocation can also be filed even after the term of a patent has expired.

Any patent granted can be revoked on a petition of interested person on the following grounds:

1. If the invention claimed in any claim of the complete specification was already claimed in another patent already granted in India.

2. If a patent is granted to a person not entitled to apply for a patent.

3. If a patent is wrongfully obtained by contravening the rights of the petitioner or any other person.

4. If any of the claim of complete specifications is not an invention under the Act.

5. If the invention is not new, or if it is publicly known or used in India before the priority date of the claim or published in India before the claims was made.

6. If the invention is not useful or sufficiently described.

7. If the claims are not properly defined or claimed invention is not patentable.

8. If patent obtained by false suggestion or representation.

9. If claimed invention was secretly used before the priority date.

10. If invention falls under non-compliance of secrecy discretion.

11. If specification wrongly mentioned or not disclosing geographical origin.

12. If invention is anticipated by traditional knowledge.

Revocation is used to invalidate patents, especially when a patent under question is damaging one's business interest it is referred as an offensive weapon. A revocation petition relies on as many grounds of revocation as possible to demolish the claims of the patent. Revocation proceedings have resulted in out of court settlements between the mega co-operations, as a result many strategic alliances, joint ventures, cooperative research programmes are resulted. Only recently Sir J.C. Bose, a pioneering Indian plant physiologist and inventor of several electronic instruments was credited with the invention of the telegraph, rather than Marconi, as we had learned since childhood. Some case studies on revocation of drug patents are given in Box 3.

Box 3
Revocation of drug patents

1. In May 2010, the Chennai Patent Office had revoked Swiss firm Roche's patent on HIV drug valganciclovir (brand name Valcyte) following post-grant applications by patient groups. Valcyte was then priced at Rs. 1,023 per tablet, while the generic competitors were priced at Rs. 245 per tablet.

2. Delhi Patent Office revoked a patent granted in 2007 to US drugmaker Pfizer for sunitinib (brand name Sutent). A drug used to treat a type of kidney cancer, which had global sales of US$ 1.19 billion in 2011. The ground for the revocation was a lack of inventive step, the Patent Office said. The drug costs Rs. 4,357 per mg capsule. With a patent in place, no other brand could enter the market, thereby giving Pfizer a monopoly and the drug beyond the reach of most patients. The revocation follows a post-grant opposition by Mumbai-based generic drug maker Cipla against Pfizer's patent on the drug.

---

11. Licensing

Licensing is the right granted by an owner of an asset to another to use that asset while continuing to retain ownership of that asset. It is a fairly common method of exploiting IP. This gives the licensee the right to use, but not own, the IP. A patent on a technological creation can be licensed for commercial use and exploitation. The owner of the rights will usually get payments in return for their use. Payment often takes the form of royalties, although in some cases single payment may be more appropriate. The value of these rights is a commercial agreement. Patent rights can be transferred partly or wholly by an inventor to other persons or organizations using provisions in the statute through assignments, mortgages, exclusive or non-exclusive licenses, etc. [See chapter 10].

12. Compulsory Licensing

Compulsory licensing (CL) allows governments to license third parties (that is, parties other than the patent holders) to produce and market a patented product or process without the consent of patent owners. CL is issued to prevent the abuse of patent as a monopoly and for commercial exploitation of the invention. Compulsory license can be issued after three years from the date of grant of patent, provided, a) reasonable requirements of public have not been satisfied; b) patented invention is not available to public at a reasonably affordable price or c) patented invention is not worked in India (Section 84).

Applicant's capability including risk taking, ability of the applicant to work the invention in public interest, nature of invention, time elapsed since sealing, measures taken by patentee to work the patent in India are to be taken into account by the Controller of Patents before granting CL. In case of national emergency or other circumstances of extreme urgency or public non commercial use or an establishment of a ground of anti competitive practices adopted by the patentee, the above conditions will not apply.

CL can be obtained for manufacture and export of patentable pharmaceutical product to any country having insufficient or no manufacturing capacity in the pharma sector for the concerned product to address public health problems. This section is an "enabling provision" under Section 92A.

While settling the terms and conditions of compulsory licenses, the Controller should endeavour to secure:

- that the royalty and other remuneration, if any, reserved to the patentee or other person beneficially entitled to the patent, is reasonable, having regard to the nature of the invention, the expenditure incurred by the patentee in making the invention or in developing it and obtaining a patent and keeping it in force and other relevant factors;
- that the patented invention is worked to the fullest extent by the person to whom the license is granted and with reasonable profit to him;
- that the patented articles are made available to the public at reasonably affordable prices;
- that the license granted is a non-exclusive license;
- that the right of the licensee is non-assignable;
- that the license is for the balance term of the patent unless a shorter term is consistent with public interest;
- that the license is granted with a predominant purpose of supply in the Indian market and that the licensee may also export the patented product if required;
- that in the case of semi-conductor technology, the license granted is to work the invention for public non-commercial use;
- that in case the license is granted to remedy a practice determined after judicial or administrative process to be anti-competitive, the licensee shall be permitted to export the patented product, if need be.

The Paris convention also recognized that non working of registered patents may amount to abuse of the patent system. Article 5A(2) permits members of Paris union to "take legislative measures for the grant of compulsory licenses, to prevent the abuses which might result from the exercise of exclusive rights conferred by the patent. Article 5A (3) persistent inaction on the part of the patent holder may even be remedied by forfeiture, of the patent right, but not before the expiration of two years from the grant of a first compulsory license. According to Article 5A (4) however, insufficient working shall not result in compulsory license before the expiration period of 4 years from the date of filing of patent application or 3 years from the date of grant of patent, whichever period expires last. The patent holder is allowed to justify his inactions by legitimate reasons stemming from "the existence of legal, economic, technical obstacles to exploitation, or more intensive exploitation of the patent in the country.

The TRIPs agreement describes the minimum rights that a patent owner must enjoy. But it also allows certain exceptions. A patent owner could abuse his rights, for example by failing to supply the product on the market. To deal with that possibility, the agreement says governments can issue "compulsory licenses (Article 31) for allowing a competitor to produce the product or use the process under license. But this can only be done under certain conditions aimed at safeguarding the legitimate interests of the patent-holder.

This provision is introduced to address the public health concerns of the countries having insufficient or no manufacturing capacity in the pharmaceutical sector to implement the decision of the TRIPs council on Para 6 of the Doha Declaration on TRIPs agreement and Public Health. This section lays down the conditions that are required to be fulfilled, when the compulsory licences for export purposes will be available. The compulsory license is available only for; (a) The patented pharmaceutical product; (b) Manufacture and export to any country having insufficient or no manufacturing capacity in the pharmaceutical sector; (c) The product addressing the public health problems in such country.

Table 2: Representative examples of compulsory license of drug patents in various countries (Kaur and Chaturvedi, 2015)

S.No.	Country	Year	Drug	Remarks
1.	USA	2001	Ciprofloxacin	Reduction in 54% of original price of the drug
2.	Zimbabwe	2002	Lamivudine and Ziduvudine	Reduction in price of drug helped Zimbabwe in the period of emergency for AIDS
3.	USA	2004	Latanoprost and Ritonavir	Due to threat under Bayh-Dole Act, Patent holder lowered the price of drug to affordable value
4.	Malaysia	2004	Didanosine, Zidovudine and combination of Lamivudine and Zidovudine	Two year compulsory license was issued in import the drugs from India
5.	Indonesia	2004	Lamivudine and Nevirapine	The generic version of drugs was available in very affordable price. The license was issued for government use, and it includes a royalty rate of 0.5% of the net selling value
6.	Mozambique	2004	Lamivudine, Stavudine and Nevirapine	The license was granted to local producer, Pharco Mozambique to produce fixed dose combination but the plan had to be shelved because the price of APIs was economically very high
7.	Zambia	2004	Lamivudine, Stavudine and Nevirapine	The license was granted to a local producer to produce a triple fixed dose combination

Contd...

S.No.	Country	Year	Drug	Remarks
8.	Taiwan	2004	*Tamiflu*	In 2007, Taiwan drug firms can make Tamiflu for domestic use and should use it only when there is a shortage of supply from Roche
9.	Thailand	2007	*Lopinavir and Ritonavir*	These antiretroviral drugs came in approach of public due to reduction in their price. Royalty of 0.5% was given to the patent holder
10.	Thailand	2007	*Clopidogrel*	Myocardial ischemia and cerebro-vascular accident being the most serious public health burden because of high mortality and disability loss. Its mortality rate is in top three annual ranking. So with the grant if its compulsory licence, the mortality rate got reduced
11.	Indonesia	2007	*Efavirenz*	Compulsory licensing reduced the price of drug and increased its accessibility
12.	Ecuador	2010	*Lopinavir and Ritonavir*	The patent was held by Abbott Pharmaceuticals. The term of License was the time that was left for the patent i.e. November 2014
13.	Ecuador	2010	*Ritonavir*	Till 2014, Ecuador issued nine compulsory licenses for various drugs including *Sutinib, Certolizumab, Mycofenolate sodium* etc.
14.	Cameroon	2005	ARVs such as, *Lamivudine, Nevirapin, Zidovudine*	CL issued to essential inventions for manufacture of anti HIV drugs
15.	Eretria	2005	Anti HIV/AIDS drugs	Compulsory license issued to import Anti HIV drugs
16.	Guinea	2005	Anti HIV/AIDS drugs	Compulsory license issued to import Anti HIV drugs

Box 4

Bayer Loses Cancer Drug Nexavar Compulsory License Patent to Natco

Patent regulatory authority, the Intellectual Property Appellate Board (IPAB), in 2013 upheld the country's compulsory license (CL) issued in 2012 to Hyderabad based Natco Pharma on Bayer's cancer drug Nexavar, setting an important legal precedent and paving the way for more generic companies to challenge patents. "Affordability and "access" are said to be the major reasons for the patent regulator's decision and should clear the decks for exorbitantly priced life-saving drugs to be made at a fraction of the price by generic companies. German MNC Bayer Corporation held a patent on renal cancer drug- Nexavar. The Natco Pharma, Hyderabad based company's application seeking approval to manufacture generic Nexavar through a CL was cleared by Indian Patent Office bringing down its price by 97% at Rs. 8,880 (around $175) for a 120-capsule pack for a month's therapy as compared to Bayer's price of over Rs 2.8 lakh (roughly $5,500) per 120 capsules. However, IPAB hiked the royalty to be paid to the German company to 7% from 6% announced last year by Patent office. India's first CL is seen as a prospective watershed for affordable access to patented medicines by potentially opening the way for other life-saving drugs including latest HIV drugs now patented in India and priced out of reach. The decision means that the way has been paved for CLs to be issued on other drugs, now patented in India and priced out of affordable reach, to be produced by generic companies and sold at a fraction of the price.

Box 5
First Compulsory License for HIV-AIDS Drug in Thailand

Thailand Government issued the first compulsory license for a drug, which is a generic version of Efavirenz, a drug of Merck & Co. Inc. The pioneering act has been given the State's approval with a view to lower costs of the drug that is essential for patients suffering from HIV-AIDS. In furtherance of its social awareness of the disease and in acknowledgement of the disease acquiring an epidemic status, Thailand's State owner Pharmaceutical manufacturer, Government Pharmaceutical Organization (GPO), declared that it would import generic versions of Efavirenz used as part of the antiretroviral therapy for the treatment of HIV-AIDS from India until it could gain the ability to make the same internally. The move is in furtherance to Thailand's strong stance against the disease. However, the compulsory licensing has a catch and has been accepted by Merck & Co Inc. on the understanding that the GPO producing the drug would pay 0.5 percent royalty on the sales of the generic version of Efavirenz to Merck & Co. The drug would increase the lifespan of people suffering from the disease and would enable them to do so at affordable costs.

Compulsory licenses have been granted by different countries especially for pharmaceuticals to safeguard the interest of public and affordability of drugs at reasonable prices (Table 2). Case studies have been given in Boxes 4 and 5.

13. Infringement of Patent

Infringement of a patent consists of the unauthorized making, importing, using, offering for sale or selling any patented invention within the India.

Remedies against infringement of a patented invention:

1. Interlocutory injunction: A patent owner at the start of a trial can request for an interim injunction to restrain the defendant from committing the acts complained of until the hearing of the action or further orders. Permanent injunction is given based on the merits of the case at the end of the trial.

2. Relief of damages: An award of damages focuses on the losses sustained by the claimant. A patent owner is entitled to the relief of damages as compensation to the patentee and not punishment to the infringer.

3. Account of profits: Account of profits focuses on the profits made by the defendant, without reference to the damage suffered by the claimant at the hands of the defendant. The purpose of the account is to prevent the unjust enrichment of the defendant by the use of the claimant's invention. The patent owner may also opt for the account of profits where he has to prove use of invention and the amount of profit derived from such illegal use.

Penalties

Penalties for different types of contraventions are given below:

1) Contravention of secrecy provisions relating to certain inventions (Sec. 118) – If any person fails to comply with any directions given under Sec. 35 relating to inventions relevant for defence purposes or under Sec. 39 which deals with residents not to apply for patents outside India without prior permission shall be punishable with imprisonment up to 2 years or with fine or with both.

2) Falsification of entries in register etc (Sec.119): If any person makes, or causes to be made, a false entry in any register kept under this Act, he shall be punishable with imprisonment for a term that may extend to 2 years or with fine or with both.

3) Unauthorized claim of patent rights (Sec. 120): If any person falsely represents that any article sold by him is patented in India or is the subject of an application for a patent in India, he will be punishable with fine that may extend to Rs. 1,00,000. The use of words 'patent', Patented', 'Patent applied for', 'Patent pending', 'Patent registered' without mentioning the name of the country means they are patented in India or patent applied for in India.

4) Wrongful use of words, "patent office" (Sec. 121): If any person uses on his place of business or any document issued by him or otherwise the words "patent office" or any other words which reasonably lead to the belief that his place of business is, or is officially connected with, the patent office, he will be punishable with imprisonment for a term that may extend to 6 months, or with fine, or with both.

5) Refusal or failure to supply information (Sec. 122): If any person refuses or fails to furnish information as required under section 100(5) and 146 he shall be punishable with fine, which may go up to Rs 10,00,000/-. If he furnishes false information knowingly he shall be punishable with imprisonment that may extend to 6 months or with fine or with both.

6) Practice by non-registered patent agents (Sec. 123): Any person practicing as patent agent without registering is liable to be punished with a fine of Rs. 1,00,000/- in the first offence and Rs.5,00,000/- for subsequent offence.

7) Deals with offences by companies (Sec. 124): When offence is committed by a company as well as every person in charge of and responsible to the company for the conducts of its business at the time of the commission of the offence will be deemed to be guilty and will be liable to be proceeded against and punished accordingly. Provided that nothing contained in this sub-section shall render any such person liable to any punishment if he proves that the offence was committed without his knowledge or that he exercised all due diligence to prevent the commission of such offence.

14. Appellate Board

Under section 83 of the Trade Marks Act, 1999 an appellate board has been established for making appeals to the decisions made by the Controller of Patents, Designs and Trade marks and other closely related matters. The appellate board consists of Chairman, technical members and other staff. Appeal should be made within three months from the date of the decision, order or direction of the Controller or the Central Government. All cases of appeals against any order or decision of the controller and all cases pertaining to revocation of patent other than on a counter claim in a suit for infringement and rectification of register pending before any High Court shall be transferred to the Appellate Board.

15. Bolar Provision and Parallel Import Rights

Under section 107A(a) of the Patents Act, 1970, Amendment 2005, any act of making, constructing, using, selling or importing a patented invention solely for uses reasonably related to the development and submission of information required under any law for the time being in force, in India, or in a country other than India, that regulates the manufacture, construction, use, sale or import of any product shall not be considered as a infringement of patent rights. Some countries allow manufacturers of generic drugs to use the patented inventions for development and submission of information required under law – e.g. — from public health authorities-without the patent owners permission and before the patent protection expires. The generic producers can then market their version as soon as the patent expires. This provision is called "Bolar-like provision" or "regulatory exception".

Parallel import provisions are provided in section 107A(b), which says that importation of patented products by any person from a person who is duly authorized under the law will not be considered as an infringement. Therefore it is possible to import the patented products from the licensee of the patentee in any country without the permission of the Patentee. The purpose of Parallel import is to check the abuse of patent rights and meant to control the price of patented product.

Selected reading

Chawla, H.S. and Singh, A.K (2007) Intellectual Property Rights: Patents, Plant Variety Protection and Biodiversity, Published by Intellectual Property Management Centre, G.B. Pant Univ. of Agric & Tech., Pantnagar, pp 54.

Diamond vs. Chakrabarty, 447 U.S. 303,1980.

Directive 98/44/EC of the European Parliament and of the Council of 6 July 1998 on the legal protection of biotechnological inventions, 1998 *Official J. Eur. Communities* O.J. (L 213) 13-21, http://europa.eu.int/eur-lex/pri/en/oj/dat/1998/l_213/ l_21319980730en00130021.pdf.

European Patent Office, European Patent Convention, Part II, Chapter 1: Patentability, Art. 53 – Industrial Application, http://www.european-patent-office.org/legal/epc/e/ar53.html.

http://www.wipo.int/treaties/en/

Kaur, A. and Chaturvedi, Rekha. 2015. Compulsory licensing of drugs and pharmaceuticals: Issues and Dilemna. J Intellect. Property Rights 20: 279-287

Official Journal of The EPO, Administrative Council of 16 June 1999, amending the Implementing Regulations to the European Patent Convention, http://www.european-patent-office.org/epo/pubs/oj99/7_99/index.htm.

Restaino L G, Halpern S E and Tang E.L, Patenting DNA related inventions in the European Union, United States and Japan: A trilateral approach or a study in contrast, 2003 (http://www.lawtechjournal.com/)

The Patents Act, 1970 along with The Patent rules [vide S.O. 493 (E) dated 2nd May, 2003], (Universal Law Publishing Co., Delhi), 2010

Patenting: Biotechnology and Bioinformatics

Biotechnology is the synergistic union of the biological sciences and the technologically based industrial arts. It is the utilization of biological processes for the exploitation and manipulation of living organisms or biological systems, in the development or manufacture of a product or in the technological solution to a "real-world" problem. Biotechnology involves application of technology on biological organisms *viz*. microorganisms, plants and animals and biological material of DNA and RNA but patenting laws of most of the countries were solely based on non-biological material and inventions. The general prerequisites for patentability are novelty, inventiveness and utility. Whether these apply to living organisms and biotechnology inventions as well? Patenting and protection of microorganisms, plants and animals, their manipulation with novel tools for rDNA technology which includes cloning, regeneration protocols, promoter, vector sequences, genetic transformation methods, expressed sequence tags (ESTs), molecular markers and scientific disciplines of omics and bioinformatics have become an important issue not only for its protection but also for its commercialization.

Louis Pasteur, the famous French scientist, received U.S. Patent No. 141,072 on July 22, 1873, claiming "yeast, free from organic germs of disease, as an article of manufacture. This was a process patent. With the phenomenal growth of genetic engineering in the late 1970s, the patentability of living microorganisms came in to the scene. The first patent on living organism was granted to Dr Chakrabarty in 1980 for a new micro-organism *Pseudomonas* bacterium in USA. USPTO rejected the claim on *Pseudomonas* bacterium as product of nature, but the case went to Supreme Court and decision was in favour of Chakrabarty in a landmark case, *Diamond (USPTO commissioner)* vs. *Chakrabarty (inventor)*.

The patenting of inventions in the field of biotechnology poses challenges to the applicants for patents as well as to the Patent Office. Therefore, there is an urgent need to put in place Guidelines to establish uniform and consistent practices in the examination of patent applications in the field of biotechnology and allied subjects under the Patents Act, 1970.

1.   Brief History of Patenting of Biotechnology in India

Till 2002, as per the prevailing practice in the Patent Office in India, patents were not granted for inventions relating to; (a) living entities of natural or artificial origin; (b) biological materials or other materials having replicating properties; (c) substances derived from such materials; and (d) any processes for the production of living substances/entities including nucleic acids. However, patents could be granted for processes of producing non-living substances by chemical processes, bioconversion and microbiological processes using micro-organisms or biological materials. For instance, claims for processes for the preparation of antibodies or proteins or vaccines consisting of non-living substances were allowable. However, a land mark judgement in 2002 by Calcutta High Court in Dimminaco AG v/s. Controller of Patents, Designs and Trade Marks opened the doors for grant of patent on microorganism where the final product of the claimed process contained living organisms [See Box 1].

Box 1

Patent on Microorganisms: Dimminaco AG v/s Controller of Patents, Designs and Trade Marks

The Dimminaco case was related to a process for the preparation of a live vaccine for protecting poultry against Bursitis infection. The Controller of Patents had refused the application for grant of patent on the ground that the vaccine involved processing of certain microbial substances and contained gene sequence. The Controller had decided that the said claim was not patentable because the claimed process was only a natural process devoid of any manufacturing activity and the end-product contained living material. In 2002, the Hon'ble Calcutta High Court, in its decision in 'Dimminaco AG v/s Controller of Patents, Designs and Trade Marks', opened the doors for the grant of patents to inventions where the final product of the claimed process contained living microorganisms. The court concluded that a new and useful art or process is an invention, and where the end product (even if it contains living organism) is a new article, the process leading to its manufacture is an invention.

The Hon'ble High Court held that the word "manufacture" was not defined in the statute therefore, the dictionary meaning attributed to the word in the particular trade or business can be accepted if the end product is a commercial entity. The court further held that there was no statutory bar in the patent statute to accept a manner of manufacture as patentable even if the end product contained a living organism. The court asserted that one of the most common tests was the vendibility (salable or marketable) test. The said test would be satisfied if the invention resulted in the production of some vendible item or it improved or restored.

The Biological Diversity Act, 2002 provides a mechanism for access to the genetic resources and benefit sharing accrued there from. Section 6 of the BD Act came into force on 1st July 2004, and prescribes that obtaining IPRs from the utilization of biological resources in India is subject to the approval of the National Biodiversity Authority. To facilitate this access and benefit sharing and in order to prevent any unauthorized use of the biological resources of India, in 2005 suitable amendments were made in Section 10 of the Patents Act, 1970, wherein disclosure of the source and geographical origin of the biological material was made mandatory in an application for patent when the said material is used in an invention. In addition, a declaration by the applicant regarding the required permission from the competent authority was inserted in Form 1 of the Patents Rules, 2003. Therefore, the issues related to the BD Act and those related to mandatory disclosure of the source and geographical origin constitute an essential element of examination of biotechnology related subject matters.

2. Microorganism Patents

The first patent on living organism was granted to Dr A. Chakrabarty in 1980 for a new micro-organism *Pseudomonas* bacterium in USA. USPTO rejected the claim on *Pseudomonas* bacterium as product of nature, but the case went to Supreme Court and decision was in favour of Chakrabarty in a landmark case, *Diamond (USPTO commissioner)* vs. *Chakrabarty (inventor)*. Chakrabarty's *Pseudomonas* bacterium was manipulated to contain four plasmids controlling the breakdown of hydrocarbons was "a new bacterium with markedly different characteristics from any found in nature". The Supreme Court stated that new microorganisms not found in nature were "manufacture" or "composition of matter" within the meaning of US Patent Act §101 and thus patentable. The "product of nature" objection as mentioned by USPTO for rejection of patent therefore failed and the modified organisms were held patentable. This precedent is being followed even today to define the patentability of microorganisms. Following this decision, EPO and the Japanese Patent Office (JPO) also started granting patent protection for microorganisms in 1981. A provision of EPC, Article 53(b) is relevant here which states that patents shall not be granted for "plant or animal varieties or essentially biological processes for the production of plants or animals, however the provision does not apply to "microbiological processes or the products thereof".

The microorganisms and microbiological inventions can be patented in India under Patents Act, 1970 Amendment 2002 provided the strain is new. This provision of patent protection to microorganisms has been implemented from 20th May 2003 in India Earlier the inventions on microorganisms were not patentable and this was one of the TRIPs regulations under the Article 27.3(b) that 'parties may exclude from patentability plants and animals other than microorganisms and essentially biological processes for the production of plants or animals other than non-biological and microbiological processes'. Thus one of the conditions of TRIPs regulations have been met and enforced in the country. Inventor has to deposit the new strain in any recognized international depository. Budapest Treaty is an international convention governing the recognition of microbial deposits in officially approved culture collections which was signed in Budapest in 1973 and later on amended in 1980. Because of the difficulties and on occasion of virtual impossibility of reproducing a microorganism from description in the patent specification, it is essential to deposit a strain in a culture collection centre for testing and examination by others. It obviates the need of describing a microorganism in the patent application and further samples of strains can be obtained from the depository for further working on the patent. There are 34 International depositories for deposition of microbial cultures. India signed the Budapest treaty on 17 December 2001. In India, Microbial Type Culture Collection and Gene Bank (MTCC) at Institute of Microbial Technology (IMTECH), Chandigarh is a recognized international depository for certain kinds of microorganisms. Before filing the patent application involving microorganism, inventor(s) have to deposit the strain to International Depository Authority, which will issue the number and this has to be quoted in the patent application.

3. Plant Patents

Patents have been used to support innovations, but for various reasons plant varieties and methods of agriculture have been excluded from patentability until recently, when a few countries namely USA, Japan and Australia began granting patents on plant varieties. The US Plant Patent Act (PPA), enacted in 1930 allowed granting of property rights for plant varieties of asexually propagated plants. The patent rights were extended to distinct and new asexually reproduced plants for a period of seventeen years. To boost private industry and advances in breeding technology led to the enaction of Plant Variety Protection Act (PVPA) in the 1970 in USA. The PVPA provided protection for sexual reproduction in plants including seed germination. In 1980 Diamond vs. Chakrabarty case set in motion the trend towards the legal acceptance of the commodification of germplasm. (Commodification is the process whereby an object, whether tangible, such as seed, or intangible, such as knowledge about the seed, is turned into a commodity, i.e. something that acquires an economic worth and can be bought and sold). US Supreme Court in Diamond vs. Chakrabarty case decided that microorganism should not be precluded from patentability for the objection raised by USPTO on the basis of "product of nature". The court held that a live, man- made bacterium was patentable under the PPA and the 'product of nature' objection therefore failed and the modified organisms were held patentable. The first **utility patent** on plant was given in 1985 to Tryptophan overproducer mutants of cereal crops (US Patent No. 4,642,411) referred as Hibberd case (inventors Hibberd, K.A., Anderson, Paul C. and Barker, Melanie issued to Molecular Genetics Research and Development, USA). Following the principle established in the Chakrabarty case, it was decided that normal US utility patents could be granted for other types of plants also e.g. genetically modified plants. It was affirmed by a ruling of US Supreme Court on December 10, 2001 that plant utility patents could be granted to sexually reproduced plants in an infringement lawsuit for sexually reproduced corn hybrids against J.E.M. AG Supply Inc. by

Pioneer Hi-Bred International Inc. The court held that newly developed plant breeds fall within the subject matter of 35 USC §101 and neither the PPA nor the PVPA limits the scope of its coverage (Chawla, 2005). Among transgenic plants herbicide resistant cotton, canola, soybean, etc; insect resistant potato, cotton, maize, etc. have been patented. In Japan also plant patents are allowed.

India and so many other countries do not protect plants by strict patenting system. But there is a mandate in the TRIPs agreement that plant varieties must be protected. In pursuance to the TRIPs agreement, India has enacted "Protection of Plant Varieties and Farmers' Rights" (PPV&FR) Act, 2001, a *sui generis* system of plant variety protection. This has been discussed in Chapter 11.

4. Animal Patents

The question of whether multicellular animals could be patented was examined by the USPTO in 1980s. In 1987, *Ex Parte* Allen case the key issue was the patentability of polyploid pacific coast oysters that had an extra set of chromosomes. The applicant sought to patent a method of inducing polyploidy in oysters as well as the resulting oysters as products-by-process. However, USPTO rejected the patent application on the ground of obviousness. On April 12, 1988 USPTO issued the first patent on transgenic non-human animal "Harvard Mouse" (U.S. Patent No. 4,736,866) developed by Philip Leder (Harvard University) and Timothy Stewart. The "Harvard Mouse", was created through a genetic engineering technique of microinjection. To the fertilized egg a gene known to cause breast cancer was injected and then this egg was surgically implanted in to the mother so that she may bring it to the term. The resulting transgenic mice were extremely prone to breast cancer. After initial reluctance by the EPO, European patent was issued in 1992 (Chawla, 2005). The new provisions of EPC in 1999, Rule 23c states that inventions concerning biological materials, such as DNA, microbiological process, plants, and animals are patentable only if "the technical feasibility of the invention is not confined to a particular plant or animal variety". Further the EPC has prohibited patents on plants and animals as per EPC Article 53(b) mentioned in the category of plants and on *ordre public* or morality [Article 53(a)]. EPC has stated that certain inventions are excluded from patentability whose exploitation is contrary to *ordre public* or morality, namely: processes for cloning human beings; processes for modifying the germ line genetic identity of human beings; use of human embryos for industrial or commercial purposes; and processes for modifying the genetic identity of animals which are likely to cause them suffering without any substantial medical benefit to man or animal, and also animals resulting from such processes.

Indian Patents Act, 1970 amendment 2002 has excluded from patentability under section 3(j) "plants and animals as a whole or any part thereof other than microorganisms but including seeds, varieties and species and essentially biological processes for production or propagation of plants and animals" and section 3(i)"any process for medical, surgical, curative, prophylactic (diagnostic, therapeutic), or other treatment of human beings, or any process for a similar treatment of animals to render them free of disease or to increase their economic value or that of their products". This is in pursuance to the TRIPs agreement Article 27.3(a) and (b). This in India modified animals are not patentable and there is no TRIPs binding to patent it.

Further TRIPs Article 27.2 mentions that States may exclude from patentability inventions, the prevention within their territory of the commercial exploitation of which is necessary, to protect *ordre public* or morality including to protect human, animal or plant life or health or to avoid serious prejudice to the environment, provided that such exclusion is not made merely because the exploitation is prohibited by law. Thus, human beings or their treatment procedures are neither patentable in India nor anywhere else.

Modified animals are patentable in USA, Japan, Korea, Hungary, South Africa and few other countries but not in India. Likewise Patent offices of USA, Japan and Australia grant patents on human body parts such as limbs, organs and tissues. The making of human body parts is not viewed as invention since they exist in nature but modified or isolated body parts are viewed as multi-cellular organisms and treated as such for patentability if they meet the statutory requirements.

5.   Patenting and Cloning

Cloning is the process of transferring nucleus of an adult multicellular organism's cell to an unfertilized egg of the same species. While transgenic cloning is when a particular gene is added to the nucleus of an adult organism cell before its transfer to an unfertilized egg of the same species. Dolly the first mammal sheep was created in 1997 by cloning [Box 2]. Creation of animals by cloning is patentable in some countries. However, patenting of human cloning issue varies in different countries. Japan banned human cloning in 2001 but had permitted researchers to use human embryos that were not produced by cloning. In July 2004 Japan Government Science Council has permitted limited cloning of human embryos for scientific research. Britain and South Korea also allow cloning of human embryos for therapeutic purposes. United States prohibits any kind of human embryo cloning but allows patenting of animal cloning. In the controversial issue of cloning, in Europe in July 1998, a European directive (98/44/EC) was adopted on the legal protection of biotechnology inventions.

6.   Patenting of Stem Cells

Stem cell research involving cell transplant options presents a promising opportunity to find the best treatment for patients. Human stem cell research is an important step towards developing new forms of treatment for such diseases as type 1 diabetes and Parkinson's disease, spinal cord injury, osteoporosis, liver damage, atrial valve damage and many more. Research using human embryonic stem (ES) cells could assist in developing treatments for diseases and in discovering new pharmaceuticals. However, given the limited availability of human eggs used to create human ES cells and the ethical controversies associated with the harvesting of human eggs and the destruction of human embryos, it has been suggested that part-human stem cells be used for research purposes.

Although research is required with incentive of patent protection on such matter but because of moral concerns associated with the creation of part humans using stem cells and bioethical issues, legislations have been enacted in some jurisdictions prohibiting the creation and/or use of some part-human subject matter. Where this has occurred, patents cannot be granted on such subject matter.

The stem cell patenting and ethical issues raised concluded with:

- that human ES cells should only be used if the same scientific results cannot be reached by using human adult stem cells,
- that only human ES cells should be used which come from surplus embryos (fertilized eggs) from infertility treatment and which have been collected after voluntary informed consent,
- that multipotent/coherent stem cells from embryos should not be subjected to patenting,
- methods and processes for transformation of undifferentiated stem cells into differentiated cells (lineages) for therapeutic use and the tissues developed by using research protocols should be patentable.

Patents on the human embryonic stem cells have been granted in US while in Europe the ethics of stem cells patentability is still a controversial subject of debate [Box 3]. In UK common rules are found in the Patent Act 1977 and the provisions of Directive, which address patentability, were introduced into UK law in July 2000. The new "Patent Regulations 2000" are in the Section 76 A.02 of the UK Patent Act. It states that an invention shall not be non patentable solely on the grounds that it concerns (i) a product consisting of or containing biological material; or (ii) process by which biological material is produced, processed or used. However, it then sets out the following as not being patentable inventions: a) The human body, at the simple of its formation and development, and the simple discovery of one of its elements, including the sequence or partial sequence of a gene; b) Processes for cloning human beings; c) Processes for modifying the germ line genetic identity of human beings; d) Uses of human embryos for industrial or commercial purposes; e) Processes for modifying the genetic identity of animals which are likely to cause them suffering without any substantial medical benefit to man or animal, and also animals resulting from such processes; f) Any variety of animal or plant of any essentially biological process for the production of animals or plant, not being a micro-biological or other technical process or the product of such a process.

Box 2

Roslin Institute (Edinburg) holds patent on quiescent cell populations for nuclear transfer (WO 9707669 (A1)). The patent was filed in 1996. It claimed a reconstituted animal embryo prepared by transferring the nucleus of a quiescent donor cell into a suitable recipient cell, in which the animal is an ungulate or rodent. The patent is valid till 2016. It also has patent on quiescent cell populations for nuclear transfer in the production of non-human mammals and non-human-mammalian embryos (US 6,147,276) which claimed a method of reconstituting a non-human mammalian embryo by transferring the nucleus of a quiescent diploid donor cell into a suitable enucleated recipient cell of the same species, thereby obtaining a reconstituted cell.

Box 3

European Court of Justice in 2011 banned patents based on human embryonic stem cells. Judgement was delivered in a patent case pertained to Oliver Brustle's, University of Bonn, Germany on a method for generating neurons from human embryonic cells. The Court's ruling, which cannot be appealed and applies to all 27 member states of the European Union (EU), bans patents on procedures that involve the destruction of human embryos at any stage. That includes not only procedures in which embryonic stem-cell lines are created, but also those that use previously derived cell lines. The court ignored statements in favour of such patents from the European Commission and several EU countries, including the UK, Sweden, Portugal and Ireland. There is however, a debate on whether the ruling applies to procedures that do not involve the destruction of embryos. Existing European patents involving embryonic stem cells-most of which were issued in the UK not be invalidated immediately, but lawsuits challenging individual patents will use the ruling as guidance.

---

6. Patenting in Biotechnology

6.1 Biological Compounds

Biological compounds, such as DNA, RNA and proteins, are not themselves living, but naturally occur in nature. The ability to isolate genes and produce the proteins they encode has enormous commercial impact. The availability and scope of patent protection on genes and genome-related technologies is considered vital for the survival and success of the biotechnology industry. Under U.S. Patent law, DNA sequences are considered chemical compounds by the USPTO and are patentable as compositions of matter. In its "Utility Examination Guidelines," the USPTO explained that isolated and purified DNA molecule that has the same sequence as a naturally occurring gene is different from the naturally occurring compound as it is processed through purifying steps that separate the gene from other molecules naturally associated with it and hence eligible for patent protection. If a patent application discloses only nucleic acid molecular structure for a newly discovered gene, and no utility for the claimed isolated gene, the claimed invention is not patentable. Since one of the requirements of a patent is utility.

Indian Patents Act, 1970 allows inventions on isolation for a substance like DNA, gene sequences, polypeptide sequences, vector molecules, etc. for which the guidelines for biotechnology applications for patents have been issued in 2013. For example a claim to a polynucleotide sequence that was available as part of a library before the priority date, lacks novelty, even if activity or function of the said sequence of the polynucleotide has not been previously determined. If any sequence of a polynucleotide/polypeptide from a prior art does not exactly match with the claimed sequence of polynucleotide/polypeptide, then the subject-matter of such claims cannot be said to be anticipated by the prior art sequence. However, such sequence of polynucleotide/polypeptide of the prior art would be relevant for deciding inventive step or non-patentability under relevant clauses of Section 3 of the Act. In another situation, if the claimed invention relates to a polynucleotide/polypeptide having mutation(s) in a known sequence of polynucleotide/polypeptide, which does not result in an unexpected property whatsoever, then the claimed subject-matter lacks inventive step. However, if the claimed invention having mutation in a strategic location with unexpected property, then the claim is under consideration.

In another example a recombinant DNA sequence of SEQ ID NO: X encodes human interferon $\alpha2$ polypeptide, while prior art discloses a nucleic acid sequence of SEQ ID NO: X1 encodes human interferon $\alpha1$ polypeptide. The claimed human interferon $\alpha2$ is structurally close to the prior art's human interferon $\alpha1$. However, the alleged invention can be held non-obvious, because of the fact that the claimed human interferon $\alpha2$ is thirty times more potent in its antiviral activity than its prior art analogue of $\alpha1$.

It must be remembered that the use of claimed subject-matter (e.g. a gene or a protein) disclosed in the specification should not be merely speculative, rather the said use should be specific, substantial and credible for establishing industrial applicability of the claimed subject-matter. Products such as microorganisms, nucleic acid sequences, proteins, enzymes, compounds, etc., which are directly isolated from nature, are not patentable subject-matter. However, processes of isolation of these products can be considered subject to patent eligibility requirements of Section 2 (1) (j) of the Act. The inventions relating to three-dimensional or crystal structure of a polypeptide attracts the provision of Section 3 (d) of the Act unless it is proved that such polypeptide differs significantly in the properties with regards to therapeutic efficacy.

6.2 Plant Regeneration and rDNA

Patents have been given on different aspects of plant tissue culture and biotechnology. If one considers different approaches/processes of regeneration (organogenesis, embryogenesis) in a particular species then it is possible to obtain patent on these processes provided it fulfills the requirements. The biological material such as recombinant DNA, plasmids and processes of manufacturing thereof are patentable provided they are produced by substantive human intervention. Gene sequences, DNA sequences with their disclosed functions are patentable if it fulfills the requirements of patentability. Promoter, terminator sequences have been patented because it fulfills the requirements as a gene sequence which has a particular function.

Patent (No. ZA9903893) on cloning of adult, selected plants of Eucalyptus through an *in vitro* regeneration process by somatic embryogenesis has been granted in Zimbabwe. Also, in Japan (Pat. No. JP2002191246) a method of producing cloned seedlings of Eucalyptus plants is with Toyota Motor Corp., Japan. Besides, patents on procedures of *in vitro* cloning in millets (*Panicum miliaceum, Echinochloa frumentacea*) and rice with Patent Nos. RU 2226819, RU2218755, RU2203534 and Patent Nos. US5350688, US6153813 respectively have been granted. These patents in general employ tissue culture procedures for multiplication and regeneration resulting in cloning of plants.

In rDNA technology first thing is to produce recombinant DNA molecules. These rDNA molecules are patentable which are made by the use of restriction enzymes and ligases. As you are aware Cohen and Boyer genetically engineered rDNA molecule way back in 1973, but they were also reluctant to file patent. They were ultimately forced to do the same. The story of Cohen and Boyer is given in Box 4.

Box 4

The Story of Cohen and Boyer rDNA patents

Stanley Cohen, a professor of medicine at Stanford University and Herbert Boyer, a biochemist and genetic engineer at Univ. of California worked together and perfected a technique. Boyer's team and Cohen had developed a method for introducing antibiotic carrying plasmids into certain bacteria as well as a method of isolating and cloning genes carried by the plasmids. Within four months of joining they had a breakthrough in cloning predetermined patterns of DNA and the technique of recombinant DNA was born which was published in 1973 with four authors. It has provided immensely valuable tools for genetic engineering. But in 1973, Cohen and Boyer had no interest in patenting the method.

Niels Reimers, founder of Stanford University's technology commercialization program in 1970 recalls that when he learnt of the published paper, he recognized the potential of this work and kept on persuading that the methodology should be patented. At that time US laws allowed one year grace period between the dates of publication and filing of a patent application. Reimers eventually succeeded in his persuasion and the patent application was filed just one week before the deadline on 4th Nov., 1974 with Cohen and Boyer as inventors. Their two coauthors were not included as they did not fulfill the legal requirements for being named as inventors. If a patent was issued, it was to be assigned to Stanford University. In 1976, Boyer, cofounded Genentech with Robert A. Swanson, a venture capitalist.

The original 1974 patent application had claimed both the process of making recombinant DNA and any products that resulted from using that product. The application was subsequently divided into a process patent application and two divisional product patent applications (One each for recombinant DNA products produced in prokaryotic cells and eukaryotic cells. The original patent application was abandoned, but the subsequent three patent applications claimed priority on the basis of original application. 6 years after the 1974 patent application, the first patent titled "Process for producing biologically functional chimeras" (US patent No. 4,237,224) was granted on 2nd Dec., 1980; the second patent titled "Biological functional molecular chimeras" (US patent No. 4,468,464) was granted on 28 Aug., 1984, and the third patent also titled "Biological functional molecular chimeras" (US patent No. 4,740,470) was granted on 26 April, 1988. Under the US patent law all three patents have expired on 2nd Dec., 1997 because they had claimed priority of invention on the basis of their 1974 patent application. Stanford University and Univ. of California had made US$255 million in licensing revenues from the patents granted to a total of 468 companies on behalf of both the universities. A total of 2442 known products were developed from the patented technology that included drugs to mitigate the effects of heart disease, anemia, cancer, HIV-AIDS, diabetes, etc. Commercial recombinant DNA products developed by the licensees generated over US$35 billion in sales during the life span of the patents.

The grant of patent did not have an easy passage for production of rDNA molecules. It took almost six years for the first patent to be granted. At that time it was assumed that inventions related to manipulation of living matter were not patentable. Decision in the case of Chakarabarti v. Diamond in the US Supreme court on 16 June, 1980 made the patenting of life forms patentable. It clarified that for patenting, "the relevant distinction was not between living and inanimate things but between products of nature, whether living or not and human made inventions. This finally cleared the Cohen and Boyer patent on 2nd Dec., 1980.

Primer sequences which have been identified as markers can get patent protection. Few examples of primer sequences which have been granted Indian patents are: Oligonucleotide primers for detection of white spot syndrome virus affecting shrimp (DBT), DNA markers for assessing seed purity and method of using DNA sequences for assessing seed purity (CSIR), A process for detection of parental and hybrid lines of rice by molecular markers (J.K. Industries), etc.

Enigma Diagnostics Ltd, has been issued European Pat. No. EP1044283, in 2009 which covers the use of endonuclease-based reporting in PCR and other nucleic acid amplification technologies. The technology involves the use of a thermostable endonuclease to cut a probe that recognizes a specific sequence in the target amplification product. The system may be applied to a wide variety of molecular diagnostic applications including the rapid detection of viral and bacterial pathogens. Enigma Diagnostics has developed products that completely automate the PCR process including sample extraction, amplification and analysis for application at the point-of-care by non-laboratory operators.

USPTO in 2007 has issued the first patent (U.S. Patent No 7,217,807) to Rosetta Genomics Ltd. (ROSG) supporting the microRNA based diagnostics and therapeutics. It covers in its first ever patent the composition of matter directed at a specific micro RNA gene found in the Human Immunodeficiency Virus and it is also the first patent issued relating to a human or viral micro RNA gene.

Sirna Therapeutics Inc has been granted two major patents on its fundamental RNA interference (RNAi) technology in 2005. The first patent (UK Pat No. 2397818) titled RNA interference mediated inhibition of gene expression using chemically modified synthetic short interfering nucleic acid (siNA) The chemical modifications are essential for the stability, potency and action of siRNA therapeutics as unmodified siRNAs degrade rapidly *in vivo*. The second patent (UK Pat No. 2396864) titled RNA interference mediated inhibition of vascular endothelial growth factor and vascular endothelial growth factor receptor gene expression using siRNA covers chemically modified siRNAs targeting a receptor of VEGF. This is the first patent issued covering the broad claims for a siRNA used against a drug target.

Transformation approaches in a particular plant species have also been patented. Few examples of patents on *Agrobacterium* mediated transformation approaches have been given in Table 3.1. Controversies on GM patents are given in Box 5.

An EST is part of a sequence from a cDNA molecule of expressed gene, therefore, it can be used to identify and locate an expressed gene. The patenting of ESTs has proved to be controversial since NIH, USA first filed patent applications on a large number of ESTs in 1991 and 1992 (AIPPI, 1992). In early 2001, the USPTO published its new "Utility Examination Guidelines" (Anonymous, 2001) which re-affirmed that ESTs are patentable subject matter, if an EST meets the statutory requirement on utility, novelty, non-obviousness and enablement. Nevertheless, a mere assertion of the utility of an EST as a probe without further disclosure of its specific function is considered not enough by USPTO to satisfy the utility and enablement requirements.

7. Bioinformatics and Patenting

Bioinformatics is a scientific discipline that encompasses all the aspects of biological information acquisition, processing, storage, distribution, analysis and interpretation. For bioinformatics the patent offices have created separate units. EPO has a separate set of examiners from the computer science and biotechnology directorates. USPTO has an entire art unit (Group Art Unit1631) - equivalent to an EPO Directorate (Maschio and Kowalski, 2001). There are three basic types of inventions on bioinformatics, which can seek patent protection.

7.1 The Tools of Bioinformatics

Computer software is one of the central tools of bioinformatics and the way in which it is treated by the patent offices varies in different parts of the world. In the USA, as early as 1969 the transformation of a computer by a computer program (using electronic signals) was recognized as patentable subject (*in re Bernhart*) (Maschio and Kowalski, 2001). USPTO in 1996 issued Examination Guidelines for Computer Related Applications. Generally under these guidelines, even though a claim contains a mathematical algorithm

Table 3.1: Patents on *Agrobacterium* mediated transformation

Patent Title	Patent Holder	Patent No.; Year	Remarks
Process for the incorporation of foreign DNA into the genome of dicotyledonous plants using stable cointegrate plasmids	Schilperoort & Hille	US 4,693,976; 1987	Process for the incorporation of foreign DNA into the genome of dicotyledonous plants using co-integrate plasmid pAL969 which is composed of plasmid R772 and pTiB6 with foreign gene incorporated in the T-region of Ti component.
A process for the incorporation of foreign DNA into the genome of dicoty ledonous plants; a process for the production of *A. tumefaciens* bacteria	Schilperoort & Hille	EP 120 515 B₁; 1990	Same as patent No. US 4,693,976 with a claim that co-integrated plasmid with foreign DNA between the 23 bp of the wild type T-region.
Transformation of tomato	Japan Tobacco Inc. (Japan)	JP4222527; 1992	*Agrobacterium* mediated trans formation in tomato with a tumor inducing plasmid pTiBo542 resulting in little risk of teratogeny.
Method of transforming monocotyledon by using scutellum of immature embryo.	Japan Tobacco Inc. (Japan)	EP0672752; 1995	Use of scutellum of an immature embryo of a monocot for *Agrobacterium* mediated transformation and the ability to regenerate normal plants after transformation.
Transformation of *indica* rice	Japan Tobacco Inc. (Japan)	EP0897013; 1998	Transformation in *indica* rice of an immature germ cell and selection in a specific culture medium with different concentrations of salts, growth regulators and carbohydrates
Agrobacterium-mediated transformation of plants	Nunhems Zaden BV (Netherlands)	US6,323,396; 2001	Enhanced transformation in cereals, vegetables and oilseeds by using auxotrophic selective medium for methionine and cysteine, or adenine and histidine.
An improved efficiency Agrobacterium-mediated plant transformation method	Monsanto, USA	US2001054186; 2001	*Agrobacterium* mediated transformation on wheat, rice, corn and soybean related to plant explant tissue preparation system with reduced moisture and weight followed by transformation resulting enhanced regeneration of fertile plants.
Method to enhance *Agrobacterium*-mediated transformation of plants.	Regents of the Univ. of Minnesota (Minneapolis)	US6,759,573; 2004	Method to enhance *Agrobacterium* mediated transformation by 10% using cysteine, glutathione, sodium thiosulfate, or dithiothreitol in soybean.
Method for super rapid transformation of monocotyledon	Nat. Inst of Agrobiological Sciences (Japan)	AU775233 B2; 2004	A method for transforming a monocot by treatment of intact seed with *Agrobacterium* containing a recombinant gene of interest.
Method of elevating transformation efficiency in plant by adding copper ion.	Japan Tobacco Inc. (Japan)	EP1669444; 2006	Transformation ability of monocot plants (maize and rice) and use of copper enriched medium during regeneration of a plant from a de-differentiated plant cell.

Box 5

The Story of Cohen and Boyer rDNA patents

1. European patent on genes that give soybean resistance to Roundup herbicide

A decision by the Europeans Court of Justice on a DNA patent held by global seed company, Monsanto has caused a stir in the biotechnology industry, with concerns that the ruling could limit the protection companies enjoy on their European patents. Since 1996, Monsanto has held a European patent on genes that give soybeans resistance to the company's Roundup herbicide, specifically the active ingredient glyphosate. But the firm has not managed to obtain a patent in Argentina, where soybean crops (known as Roundup Ready) expressing the glyphosate-resistance genes can be cultivated without a licensing agreement. Argentinean growers are exporting soya meal harvested and processed from these crops to Europe, especially the Netherlands. In an attempt to recoup payments it has not yet managed to get from Argentinian growers, Monsanto had sued importers such as Cefetra, based in Rotterdam, the Netherlands, to try to prevent this practice, claiming that the imported soya meal contained the DNA sequence that it had patent protection for in Europe. The European Court of justice based in Luxembourg ruled on 6 July 2010 that Monsanto could not bar imports of the soya meal. It argued citing the fact that the DNA in the soya meal was not performing the function for which Monsanto had gained patent protection in the first place. The ruling is being viewed as the first test of the EU'S biotechnology directive, passed in 1998, which set down policy on what kind of genetic material was patentable, and on what protection that patent enjoyed.

2. Syngenta has been granted Terminator potato patents in Australia and Russia in 2007 and has also applied for similar patents in Europe, Brazil, Canada, China, Egypt and Poland. Although in 2000 the United Nations Convention on Biological Diversity (CBD) recommended that governments should not field-test or commercialize genetic seed sterilization technologies, thus creating a de-facto international moratorium.

3. The US government has granted exclusive rights for a genetically modified (GM) Papaya, developed from papaya native to Thailand, to a team of American Researchers and Cornell University's research arm in 2007. USPTO issued patent protection to Cornell Research Foundation Inc. over its invention of GM papaya that was developed from khaek dam and kheak nual Thai native papaya varieties. The patent covers the genetic pattern and all scientific methods related to the production of novel papaya variety that is resistant to papaya ringspot virus.

4. Patented Seeds can't be Reproduced by Farmers, says US Apex Court

The US Supreme Court has ruled in favour of Monsanto over an Indiana farmer, Verman Bowman accused of having pirated the genetically-modified crops developed by the agribusiness giant Monsanto. The Court's unanimous decision focused specifically on seed production, but experts say it may also have implications on intellectual property law in medicine, biotechnology and software. The nine justices ruled that laws limiting patents "do not permit a farmer to reproduce patented seeds through planting and harvesting without the patent holder's permission." The crux of the argument was over 'patent exhaustion' which states that, after a patented item has been sold, the purchaser has "a right to use or resell that article," Justice Elena Kagan explained in the court's 10-page decision. "Such a sale, however, does not allow the purchaser to make new copies of the patented invention," she added. In a lawsuit filed in 2007, Monsanto had accused Vernon Hugh Bowman, a farmer, of infringing on its intellectual property rights by replanting, cultivating and selling herbicide-resistant soyabean seeds it spent more than a decade developing. The patented seed, which allows farmers to aerially spray Monsanto-made Roundup herbicide over their entire fields, was invented in 1996 and is now grown by more than 90 per cent of the 2,75,000 US soyabean farmers. The farmer said he had respected his contract with Monsanto and purchased new Roundup Ready seeds each year for his first planting. But he said hard times forced him to purchase a cheaper mixture of seeds from a grain elevator starting in 1999, which he used for his second planting. The mixture included Roundup Ready soyabeans, which Bowman was able to isolate and replant from 2000 to 2007.

or an abstract idea, if the claim is limited to a practical application in the technological arts it might be statutory and thus have patentable utility under §101. In 2007 The Supreme Court of USA in 7–1 decision ruled in an altercation between Microsoft and AT&T that software patents code' shipped around the world to foreign manufacturers by Microsoft on 'golden master discs' is in fact a 'blueprint' and not a 'component' of the invention. Software in abstract is not patentable, when it is simply a set of instructions detached from any medium – an 'idea without physical embodiment'. The ruling further suggests that copies of a code that can be downloaded or copied from a CD-ROM drive and installed on to a computer do qualify as patentable.

In Europe computer software until very recently has been considered non-patentable. EPC disqualifies computer programs from patentability as such under Article 52(2). Also excluded are aesthetic creations, discoveries, scientific theories, mathematical methods and other activities that are essentially non-technical in character. Despite this applicants have been able to obtain patents covering computer programs from the EPO by not claiming computer programs 'as such' which is in the exclusion list but claiming in a technical context. The computer programs are patentable as long as they are technical in nature.

In India Computer Related Inventions (CRI) guidelines were issued by Patent Office in 2015. Various terms defined in The Information Technology Act, 2000 are as:

The term "computer" as "any electronic, magnetic, optical or other high-speed data processing device or system which performs logical, arithmetic, and memory functions by manipulations of electronic, magnetic or optical impulses, and includes all input, output, processing, storage, computer software, or communication facilities which are connected or related to the computer in a computer system or computer network".

The term "computer network" as "the interconnection of one or more computers through – (i) the use of satellite, microwave, terrestrial line or other communication media; and (ii) terminals or a complex consisting of two or more interconnected computers whether or not the interconnection is continuously maintained".

The term "computer system" as "a device or collection of devices, including input and output support devices and excluding calculators which are not programmable and capable of being used in conjunction with external files, which contain computer programmes, electronic instructions, input data and output data, that performs logic, arithmetic, data storage and retrieval, communication control and other functions".

The term "data" as "a representation of information, knowledge, facts, concepts or instructions which are being prepared or have been prepared in a formalised manner, and is intended to be processed, is being processed or has been processed in a computer system or computer network, and may be in any form (including computer printouts, magnetic or optical storage media, punched cards, punched tapes) or stored internally in the memory of the computer".

Computer programme has been defined in the Copyright Act 1957 under Section 2(ffc) as "computer programme means a set of instructions expressed in words, codes, schemes or in any other form, including a machine readable medium, capable of causing a computer to perform a particular task or achieve a particular result".

The term "software" is not defined in Indian statutes and hence, for interpretation of this term, the general dictionary meaning is being used. The "software" as "the programs, etc. used to operate a computer".

The computer programme *per se* is excluded from patentability under section 3(k) apart from mathematical or business method and algorithm under The Patents Act, 1970. Claims which are directed towards computer programmes *per se* are excluded from patentability, like (i) claims directed at computer programmes/set of instructions/Routines and/or Sub-routines written in a specific language (ii) claims directed at "computer programme products"/"Storage Medium having instructions"/"Database" /"Computer Memory with instruction" i.e. computer programmes *per se* stored in a computer readable medium.

Therefore, if a computer programme is not claimed by "in itself" rather, it has been claimed in such manner so as to establish industrial applicability of the invention and fulfills all other criterion of patentability, the patent should not be denied. In such a scenario, the claims in question shall have to be considered taking in to account whole of the claims.

For being considered computer related invention as patentable, the subject matter should involve either — a novel hardware, or — a novel hardware with a novel computer programme, or — a novel computer programme with a known hardware which goes beyond the normal interaction with such hardware and affects a change in the functionality and/or performance of the existing hardware. A computer program, when running on or loaded into a computer, going beyond the "normal" physical interactions between the software and the hardware on which it is run, and is capable of bringing further technical effect may not be considered as exclusion under these provisions.

There are indicators to determine technical advancement for examining CRI applications. The examiner shall confirm that the claims have the requisite technical advancement. The following questions are looked by the examiner while determining the technical advancement of the inventions concerning CRIs:

(i)   whether the claimed technical feature has a technical contribution on a process which is carried on outside the computer;

(ii)   whether the claimed technical feature operates at the level of the architecture of the computer;

(iii)   whether the technical contribution is by way of change in the hardware or the functionality of hardware.

(iv)   whether the claimed technical contribution results in the computer being made to operate in a new way;

(v)   in case of a computer programme linked with hardware, whether the programme makes the computer a better computer in the sense of running more efficiently and effectively as a computer;

(vi)   whether the change in the hardware or the functionality of hardware amounts to technical advancement.

If answer to any of the above questions is in affirmative, the invention may not be considered as exclusion under section 3(k) of the Patents Act, 1970.

7.2   The Methods of Bioinformatics

A second development in bioinformatics is the move towards the patenting of business methods. This is especially pertinent because classical biotechnology claims, e.g. methods for generating a tangible such as RNA, DNA or protein might not provide adequate protection for the true product of bioinformatics – information. In the USA business methods are patentable subject matter. By contrast, the patenting of business methods is amongst the exclusions found in Article 52(2) EPC, in other words, they are non-patentable 'as such' under the EPC. A patentable business method (or computer program) at least as far as the EPO is concerned, must have technical elements - for example it must be at least partly computer implemented. A biological assay that involves bioinformatics need not be claimed as a conventional biological method but a biological assay that involves bioinformatics which can be claimed as a computer implemented procedure in the same style as a business method to claim the processing of data to produce a result and this type of claim might be desirable to cover the activities of customers of bioinformatics processes.

7.3   The Product of Bioinformatics

Bioinformatics produces information. In Europe however, information as such is non-patentable under EPC Article 52(2) because of its abstract nature.

However, the EPO has allowed claims directed to data in two well-known decisions of the Technical Boards of Appeal T1494/97 and T163/85 (BBC), dating from 1990 due to the technical content. It was structured in such a way that it controlled the apparatus used to interpret the data. In the USA, claims have been obtained to business methods and to methods in which the resulting product is information. Subtle differences in claim language can mean the difference between allowed subject matter and disallowed subject matter, and between claiming the invention and not. For example, in the USA, a claim to a computer readable medium with sequence data on it is considered to be non-statutory descriptive matter, however, a claim to a software programme on a disk might be statutory. The latter lies in the technological arts because software programmes are technological; the former, however, merely relates to information on a medium. The applicant must therefore ensure that, if information is to be claimed, it is claimed such as to make it technological in nature. For example, nucleic acid and protein sequence data, which is a primary data that lack any annotation is non patentable (Table 3.2). However, elements of information of this type can be combined with other sources of data to provide useful further information, which can be termed secondary information, about the function of a gene or a polypeptide. It is knowledge of function that allows us to do something useful. This information is not abstract but technical and genetic inventions that concern diagnosis of diseases, therapy, biotechnology, genetic engineering and many other established technical fields are based on an element of knowledge of gene function. Data can be technical if they provide functional information of any useful sort.

Table 3.2: Patents on bioinformatics

Patent Title	Patent Holder	Patent No; Year	Remarks
Method and apparatus for providing a bioinformatics database	Affymetrix, Inc. (Santa Clara, CA)	US 6,882,742; 2005	Organization of information relating to polymer probe array chips including oligonucleotide array chips.
Methods and systems for analyzing complex biological systems	Icoria, Inc. (Research Triangle Park, NC)	US 6,873,914; 2005	Organization of complex and disparate data into coherent data sets to serve as models for biological applications in the agricultural, pharmaceutical, forensic, and biotechnology industries.
Production and preprocessing system for data mining	Hitachi, Ltd. (Tokyo, JP)	US 6,868,423; 2004	Means capable of solving trouble in managing data formats and procedures and capable of carrying out advanced preprocessing more intuitively.
Method and system for detecting near identities in large DNA databases	ZymoGenetics, Inc. (Seattle, WA)	US 6,775,622; 2004	Comparison of DNA sequences in a large DNA database, clustering and assembling ESTs into the cDNAs, mapping assembled ESTs onto genomic sequence, mapping cDNAs onto genomic sequences and locating alternately spliced cDNAs.

Selected Reading

AIPPI Report, Question Q 150: *Patentablity Reuirement and Scope of Protection of Expressed Sequence Tags* (ESTs), *single Nuckeotide Polymorphisms* (SNPs) *and Entire Genomes,* http://www.aippi. org/repotrs/q150/gr-150-e-questions.htm.

Anonymous, 2001. 66 Fed. Reg. 1092-99.

Chawla, H.S. and Singh, A.K. 2007. Intellectual Property Rights: Patents, Plant Variety Protection and Biodiversity, Published by Intellectual Property Management Centre, G.B. Pant Univ. of Agric & Tech., Pantnagar, pp. 54.

Chawla, H.S. 2005. Patenting of biological material and biotechnology. *J Intellectual Property Rights* 10: 44-51.

Diamond vs. Chakrabarty, 447 U.S. 303, 1980.

Directive 98/44/EC of the European Parliament and of the Council of 6 July 1998 on the legal protection of biotechnological inventions, 1998 *Official J. Eur. Communities* O.J. (L 213) 13-21, http://europa.eu.int/eur-lex/pri/en/oj/dat/1998/l_213/l_21319980730en00130021.pdf.

European Patent Office, European Patent Convention, Art. 1, Oct. 5, 1973, http://www.european-patent-office.org/legal/epc/e/ar1.html

European Patent Office, Implementing Regulations to the Convention on the Grant of European Patents, Oct. 5, 1973, Rule 23c, http://www.european-patent-office.org/legal/epc/e/r23c. htm#R23c

http://www.ipindia.nic.in/ipr/patent/ipwatch.htm Controller General of Patents, Designs & Trade Marks launches Guidelines for Examination of Biotechnology Applications for Patents, 2013

http://www.ipindia.nic.in/ipr/patent/ipwatch.htm Guidelines for Examination of Computer Related Invention, 2015.

Maschio T.A. and Kowalski T.J. 2001. Bioinformatics a patenting view. Trends in Biotechnology 19: 334-339.

Ninawe, A.S. Intellectual property rights in biotechnology. *In*: Protecting intellectual property in life sciences (Eds Dominic Keating, Abha Agnihotri and Ajit Varma), Amity University Press, Noida, India, pp. 152-161

Official Journal of The EPO, Administrative Council of 16 June 1999, amending the Implementing Regulations to the European Patent Convention, http://www.european-patent-office.org/ epo/pubs/oj99/7_99/index.htm.

Restaino L G, Halpern S E and Tang E.L, Patenting DNA related inventions in the European Union, United States and Japan: A trilateral approach or a study in contrast, 2003 (http:// www.lawtechjournal.com/).

The Patents Act, 1970. along with The Patent rules [vide S.O. 493 (E) dated 2nd May, 2003], (Universal Law Publishing Co., Delhi), 2010.

Copyright

Copyright is a form of intellectual property protection granted to the creators of original works of literary, dramatic, musical, artistic, paintings, sculptures, architecture, software, maps, cinematograph films and sound recordings, drawings and certain other intellectual works. Unlike the case of patents, copyright protects only the form of expression of ideas and not the ideas themselves. There is no copyright in an idea. The copyright law protects creativity in the choice and arrangement of words, musical notes, colours, shapes and so on. Copyright was created to provide protection to composers, writers, authors and artists to protect their original works against those who copy. The owner of a registered copyright enjoys the ability of blocking the unauthorized copying or public performance of a work protected by copyright and provides the power to exploit the same for the period prescribed under the Act. Copyright protection is also granted for question papers set for examinations, research theses and dissertations prepared by students, compilation of a book on house hold account, a book of scientific questions and answers, questionnaire for collecting statistical information's, lecture notes, course materials, research reports, laboratory notebooks, etc. which comes under the class of literary works.

Copyright law protects creative expression of an idea but not fact, concept, idea, system or method of process or operation. At the same time copyright does not ordinarily protect titles by themselves or names, short word combinations, slogans, short phrases, methods, plots or factual information. Copyright does not protect ideas or concepts. Copyright law protects expression of an idea which may be found in product design, written expression, traditional artistic works, and other original works. To get the protection of copyright a work must be original. Question arises what is meant by work? A work means a literary, dramatic, musical or artistic work, a cinematograph film, or a sound recording. While Work of joint authorship" means a work produced by the collaboration of two or more authors in which the contribution of one author is not distinct from the contribution of the other author or authors.

1. Classes of Works

Copyright in the following classes of works are given which have been defined as:

Original literary, dramatic, musical and artistic works: An artistic work means a painting, a sculpture, a drawing (including a diagram, map, chart or plan), an engraving or a photograph, whether or not any such work possesses artistic quality. A dramatic work includes any piece of recitation, choreographic work, entertainment in dumb show, the scenic arrangement or acting form of which is fixed in writing or otherwise:

   i. In order to qualify for copyright protection, choreography (a form of dramatic work) must be reduced to writing usually in the form of some notations and notes; ii) scenic arrangement fixed in writing or representations of it in drawings, costumes worn by actors represented in the form of drawing are treated as artistic works in the Copyright law.

   A "Musical work" means a work consisting of music and includes any graphical notation of such work but does not include any words or any action intended to be sung, spoken or performed with the music. A musical work need not be written down to enjoy copyright protection. It also includes a work of architecture and any other work of artistic craftsmanship.

   ii. Cinematograph films: "Cinematograph film" means any work of visual recording on any medium produced through a process from which a moving image may be produced by any means and includes a sound recording accompanying such visual recording and "cinematograph" shall be construed as including any work produced by any process analogous to cinematography including video films. A cinematograph film may be a live performance like sport event, public function, or dramatic or musical performance or it may be based on the cinematograph version of a literary or a dramatic work.

   iii. Sound recordings: "Sound recording" means a recording of sounds from which sounds may be produced regardless of the medium on which such recording is made or the method by which the sounds are produced. A phonogram and a CD-ROM are sound recordings.

Besides there are terms like "Government work" which means a work is made or published by or under the direction or control of the government or any department of the government; any legislature in India; and any court, tribunal or other judicial authority in India. "Indian work" under Copyright Act 1957 means a literary, dramatic or musical work, the author of which is a citizen of India; or which is first published in India; or the author of which, in the case of an unpublished work is, at the time of the making of the work, a citizen of India.

Thus, two authors can independently get a copyright for an identical work. To secure copyright for the product it is necessary that labour, skill, and capital should be expended sufficiently to impart to the quality or feature that differentiates it from its raw material. Copyright subsists in the original adaptation of a literary work. Adaptation means the conversion of the work into a dramatic work by way of performance in public or otherwise; any abridgement of the work or any version of the work where the meaning or story is conveyed wholly by means of picture in a form suitable for reproduction in a book, newspaper, magazine or periodical. Collective works are compilations, which include encyclopedia, dictionaries, yearbooks, newspaper, review or magazine where parts of works of different authors are incorporated.

Copyright ensures certain minimum safeguards of the rights of authors over their creations, thereby protecting and rewarding creativity.

A work is protected by a copyright at the moment it is created in a tangible form (written copy, recorded music, filmed movie, digital data saved on a computer disk). Meaning thereby, that copyright comes into existence as soon as a work is created and no formality is required to be completed for acquiring or registering copyright. Registration of the work is however highly recommended because such registration is helpful in an infringement suit. As per the Copyright Act, the register of copyrights (where the details of the work are entered on registration) is a *prima-facie* evidence of the particulars entered therein. India is a member of both Berne and Universal Conventions and Indian law extends protection to all copyrighted works originating from any of the convention countries.

2. Conventions and Treaties on Copyright

Various international conventions and treaties on copyright exist. India is a member state of the following international conventions on copyright and neighboring rights.

2.1 The Berne Convention for the Protection of Literary and Artistic Works

Berne Convention was adopted on September 9, 1886. The Berne Convention is the oldest international treaty in the field of copyright for the protection of literary and artistic works. The Berne Convention has been revised several times in order to improve the international system of protection which the Convention provides. The first major revision took place in Berlin in 1908, and this was followed by the revisions in Rome in 1928, in Brussels in 1948, in Stockholm in 1967 and in Paris in 1971. The goals of the Berne Convention provided the basis for mutual recognition of copyright between sovereign nations and promoted the development of international norms in copyright protection. It was to help the nationals of its member states to obtain international protection of their rights to control, and receive payment for, the use of their creative works such as novels, short stories, poems, plays, songs, musicals, drawing, paintings, sculptures, architectural works, etc. The Convention rests on three basic principles.

  i. There is the principle of "national treatment", according to which works originating in one of the member States are to be given the same protection in each of the member States as these grant to works of their own nationals.

 ii. There is automatic protection, according to which such national treatment is not dependent on any formality.

iii. There is independence of protection, according to which enjoyment and exercise of the rights granted is independent of the existence of protection in the country of origin of the work.

Authors of works are protected, in respect of both their unpublished or published works. The exclusive rights granted to authors under the Convention include the right of translation, the right of reproduction in any manner or form, which includes any sound or visual recording, the right to perform dramatic, dramatico-musical and musical works, the right to broadcast and communicate to the public, by wire, rebroadcasting or of public recitation, the right to make adaptations, arrangements or other alterations of a work and the right to make cinematographic adaptations and reproductions of a work.

The latest (1971) Paris Act of the Berne Convention recognizes a special right in favor of developing countries with respect to unpublished work where the identity of author is unknown especially in case of folklore, possibility of granting non-exclusive and non-transferable compulsory licenses in respect of (i) translation for the purpose of teaching, scholarship or research, and (ii) reproduction for use in connection with systematic instructional activities, of works protected under the Convention. These licenses may be granted, after the expiry of certain time limits and after compliance with certain procedural steps, by the competent authority of the developing country concerned. The Berne Convention is administered by the World Intellectual Property Organization (WIPO).

2.2 Special Conventions in the Field of Related Rights: The International Convention for the Protection of Performers, Producers of Phonograms and Broadcasting Organizations "The Rome Convention"

Several international conventions on related rights are administered by WIPO. A Diplomatic conference on October 26, 1961, in Rome adopted finally the text of International Convention for the Protection of Performers, Producers of Phonograms and Broadcasting Organizations ("the Rome Convention"). The Rome Convention incorporated new categories of rights, which were referred to as "neighbouring rights", and they included broadcasts, phonograms, and performances. With the adoption of the TRIPs Agreement, the term neighbouring rights has been largely supplanted by the term "related rights" although both terms are still used. They considered the possibility that the performers, producers of phonograms and broadcasting organizations of a country would enjoy international protection even when the literary and artistic works they used might be denied protection in that country because it was not party to at least one of the major international copyright conventions. The Rome Convention therefore provides that in order to become a party to the Convention a State must not only be a member of the United Nations, but also a member of the Berne Union or party to the Universal Copyright Convention.

Besides the Rome Convention of 1961, two other international instruments have been drawn up with regard to certain related rights. These are the Convention for the Protection of Producers of Phonograms against Unauthorized Duplication of Their Phonograms, concluded in Geneva in October 1971 and generally referred to as "The Phonograms Convention" and the Convention Relating to the Distribution of Programme-Carrying Signals Transmitted by Satellite, concluded in Brussels in May 1974 and known briefly as "The Satellites Convention." These two Conventions are also within the area of related rights, and their purpose is to grant more extensive rights than those granted by the Rome Convention. Thus, it protects producers of phonograms and broadcasting organizations, respectively, against certain prejudicial acts that have been widely recognized as infringements or acts of piracy in Rome Convention.

The Phonograms Convention and the Satellites Convention to a certain extent supplement the Rome convention but their approach is different, in three main respects.

i. The Rome Convention gives the beneficiaries of related rights essentially a right to authorization or prohibition, without overlooking the safeguarding of the rights of authors while Phonograms and Satellites Conventions do not introduce private rights but rather leave the Contracting States free to choose the legal means of preventing or repressing acts of piracy in that area.

ii. Rome Convention is based on the "national treatment" principle. That means the protection prescribed is only minimum protection and rights guaranteed by that Convention to performers, producers of phonograms and broadcasting organizations enjoy the same rights in countries party to the Convention as those countries grant their nationals. The Phonograms Convention does not speak of the system of "national treatment", but defines expressly the unlawful acts against which Contracting States have to provide effective protection; consequently, the States are not bound to grant foreigners protection against all acts prohibited by their national legislation for the protection of their own nationals. For instance, countries whose national legislation provides protection against the public performance of phonograms are not obliged to make this form of protection available to the producers of phonograms of other Contracting States, because the Phonograms

Convention does not itself guarantee any protection against the use in public of lawfully reproduced and distributed phonograms. This Convention places Contracting States under the obligation to take the necessary steps to prevent just one type of activity, namely the distribution of program-carrying signals by any distributor for whom the signals emitted to or passing through the satellite are not intended.

iii. In the interests of combating piracy over the widest possible area, the new international agreements were made open to all States members of the United Nations or any of the specialized organizations brought into relationship with the United Nations, or parties to the Statute of the International Court of Justice (virtually all States of the world). In contrast the Rome Convention is a "closed" Convention, its acceptance being reserved for States party to at least one of the two major international copyright conventions.

2.3 WIPO Copyright Treaty (WCT) and WIPO Performance and Phonogram Treaty (WPPT)

In the 1970s and 1980s, a number of important new technological developments took place viz. reprography, video technology, compact cassette systems facilitating "home taping," satellite broadcasting, cable television, the increase of the importance of computer programs, computer storage of works and electronic databases, etc. After the adoption of the TRIPs Agreement under the auspices of GATT, in 1996 the WIPO Diplomatic Conference on Certain Copyright and Related Rights questions adopted two treaties, the WIPO Copyright Treaty (WCT) and the WIPO Performances and Phonograms Treaty (WPPT).

2.3.1 WIPO Copyright Treaty (WCT)

It was adopted by the Diplomatic Conference on 20 December 1996. It came into force on March 6, 2002. It deals with the protection of authors of literary and artistic works, such as writing, musical works of the fine art, photographs, computer programmes and original databases and on the right of rental in a way similar to the TRIPs Agreement. The provisions of the WCT cover the issues and the rights applicable to the storage and transmission of works in digital systems, the limitations on and exceptions to rights in a digital environment, technological measures of protection and rights management information. The right of distribution may also be relevant in respect of transmissions in digital networks.

2.3.2 WIPO Performance and Phonogram Treaty (WPPT)

It was adopted in Geneva by the Diplomatic Conference on 20 December 1996. WPPT came into force on May 20, 2002. The provisions of the WPPT are related to the "digital agenda" which covers the rights applicable to storage and transmission of performances and phonograms in digital systems, limitations on and exceptions to rights in a digital environment, technological measures of protection and rights management information. The right of distribution may also be relevant in respect of transmissions in digital networks; its scope and the right of rental.

The minimum duration of protection of the rights covered by WPPT and WCT practically corresponds to the duration under the TRIPs Agreement (fifty years) rather than under the Rome Convention (20 years). Both the treaties addresses the challenges posed by today's digital technologies, in particular the dissemination of protected material over digital network such as Internet. Therefore, they are also referred as **'Internet Treaties'**. They provide an exclusive right for authors, performers and producers of phonograms to authorize the making and availability of their works, performances and phonograms, respectively, to the public, by wire or wireless means, in such a way that members of the public may access them from a place and at a time individually chosen by them. The treaties contain provisions on obligations concerning technological measures of protection and electronic rights management information, indispensable for an efficient exercise of rights in digital environment.

These two treaties of WCT and WPPT have two types of technology adjuncts:

1. Anti-circumvention provision: It requires countries to provide adequate legal measures and effective remedies against the circumvention of technological measures, such as 'encryption' to tackle the problem of hacking.

2. Right management information: It safeguards the reliability and integrity of the online marketplace by requiring countries to prohibit the deliberate alteration or detection of electronic information.

2.4 Beijing Treaty on Audiovisual Performances (2012)

The Beijing Treaty on Audiovisual Performances was adopted by the Diplomatic Conference in June, 2012. The Treaty deals with the intellectual property rights of performers in audiovisual performances. It grants performers four kinds of economic rights for their performances fixed in audiovisual fixations, such as motion pictures: (i) the right of reproduction; (ii) the right of distribution; (iii) the right of rental; and (iv) the right of making available to the public, by wire or wireless means, of any performance fixed in an audiovisual fixation. As to unfixed (live) performances, the Treaty grants performers three kinds of economic rights: (i) the right of broadcasting (except in the case of rebroadcasting); (ii) the right of communication to the public (except where the performance is a broadcast performance); and (iii) the right of fixation. The Treaty also grants performers moral rights. The term of protection must be at least 50 years. This treaty is not in force yet.

2.5 Marrakesh Treaty to Facilitate Access to Published Works for Persons Who Are Blind, Visually Impaired or Otherwise Print Disabled (2013)

The Marrakesh Treaty to Facilitate Access to Published Works for Persons who are Blind, Visually Impaired, or Otherwise Print Disabled has a clear humanitarian and social development dimension and its main goal is to create a set of mandatory limitations and exceptions for the benefit of the blind, visually impaired and otherwise print disabled (VIPs). It introduces a standard set of limitations and exceptions to copyright rules in order to permit reproduction, distribution and making available of published works in formats designed to be accessible to VIPs, and to permit exchange of these works across borders by organizations that serve those beneficiaries. Only works "in the form of text, notation and/or related illustrations, whether published or otherwise made publicly available in any media", including audio books, fall within the scope of the regime. This treaty is not in force yet.

3. Copyright Act in India

In India, during earlier days education was considered as the gift of God. First copyright law was enacted on 18th December 1847. It covered only books but not other kinds of creative expressions. British Parliament amended the Copyright Act in 1911. Accordingly, the Government of India enacted the Copyright Act, 1914, which remained in force till 1958 when the new Copyright Act, 1957 was passed by the parliament of Independent India. The current law of Copyright in India is the Copyright Act, 1957, The Copyright Rules, 1958 and the International Copyright Order, 1999. It came into effect from January 1958. The Act has been amended in 1983, 1984, 1992, 1994 and 1999 which has been amended in 1983, 1984, 1992, 1994 and 1999, to keep together the socio-economic and technological development.

4. Registration of Work

Copyright comes into existence as soon as a work is created and no formality is required to be completed for acquiring copyright. However, it is advisable to register the work. In India work is registered in the Register of Copyrights maintained in the Copyright Office of the Department of Education for which the facilities exist. Procedure for registration of a work is covered under Chapter VI of the Copyright Rules, 1958.

a. Application for registration is to be made on Form IV

b. Separate applications should be made for registration of each work

c. Each application should be accompanied by the requisite fee prescribed in the second schedule to the Rules; and

d. The application should be signed by the applicant or the advocate in whose favour a Vakalatnama or Power of Attorney has been executed.

The Power of Attorney signed by the party and accepted by the advocate should also be enclosed.

The requirements for filing the copyright application in India are:

1) Full name, address and nationality of applicant(s) and that of author

2) The year and country of first publication of the work.

3) List of countries where the work has been published and the year of publication.

4) The year and country of last publication.

5) Three copies of the work.

6) Power of Attorney.

7) In case of labels, which can be used as trade mark, clear copyright search certificate from the Trade Marks Registry?

The following Statement of Further Particulars should be submitted in triplicate along with the Application for Registration of Copyright (Form IV):

1. If the work is to be registered

   a. An original work?

   b. A translation of a work in the public domain?

   c. A translation of a work in which Copyright subsists?

   d. An adaptation of a work in the public domain?

   e. An adaptation of a work in which Copyright subsists?

2. If the work is a translation or adaptation of a work in which Copyright subsists:

   a. Title of the original work

   b. Language of the original work

   c. Name, address and nationality of the author of the original work and if the author is deceased, the date of decease

   d. Name, address and nationality of the publisher, if any, of the original work

   e. Particulars of the authorization for a translation or adaptation including the name, address and nationality of the party authorizing.

After filing application and receiving diary number the applicant should wait for a mandatory period of 30 days so that no objection is filed in the Copyright office against the claim that particular work is created by the applicant. If such objection is filed it may take another one month time to decide as to whether the work could be registered by the Registrar of Copyrights after giving an opportunity of hearing the matter from both the parties. If no objection is filed the application goes for scrutiny from the examiners. If any discrepancy is found, the applicant is given 30 days time to remove the same. Therefore, it may take 2 to 3 month's time for registration of any work in the normal course. Any person aggrieved by the final decision or order of the Registrar of Copyrights may, within three months from the date of the order or decision, appeal to the Copyright Board.

5.   Criteria for Ownership of Copyright

The following are the criteria for getting the ownership of copyright:

1. Authors who write books or compose music under this category may create a work on his own behalf are thus the owners of the copyright.

2. Authors who create work at the instance of another person for valuable consideration, in the absence of any agreement to the contrary; the person at whose instance the work is made is the owner of the copyright. Where the commissioned work is partly sub-contracted, the person who commissioned the ultimate article and is ultimately to pay for the work is the owner of the copyright.

3. Where author creates work in the course of employment then the ownership depends upon the nature of employment.

4. In the case of a government work, government shall, in the absence of any agreement to the contrary, be the first owner of the copyright therein.

5. In the case of a work made or first published by or under the direction or control of any public undertaking, such public undertaking shall, in the absence of any agreement to the contrary, be the first owner of the copyright therein.

6. Where a person (e.g, X) provides the material to another for writing a book and the latter (ghost writer – e.g. Y) writes the book on the basis of material supplied, then the latter person (Y) becomes the owner of the copyright in the book.

7. In respect of literary, dramatic or musical work where the work is made by an employee in the course of employment by the proprietor of a newspaper, magazine or similar periodical under a contract of service or apprenticeship, the said proprietor, in the absence of any agreement to the contrary, will be the first owner of the copyright in the work in relation to publication of the work in the newspaper or magazine or periodical or to its reproduction for the said purposes. The same rule applies to photographs taken or portraits drawn or on engraving or a cinematographic film made by an employee in the course of employment. But in all other respects the author shall be the first owner of the copyright in the work.

8. In case of collective works, the first owner of the copyright is the person who has collected, edited and organized the work.

6. Criteria for Nationality

The following are the criteria of nationality requirement:

1. There is no nationality requirement for subsistence of the copyright in cases where the work is first published in India.

2. If the work is first published outside India, the author must be a citizen of India at the time of publication (or if dead then at the time of death).

3. In case of unpublished work the author must be a citizen of India or domiciled in India at the time of making the work.

4. Copyright in an architectural work will subsist only if the work is located in India irrespective of the nationality of the author.

7. Duration of Copyright

Copyrights for different aspects have different time periods of protection as given in Table 1.

Table 1: Duration of copyright for different works

Type of work	Duration
Literary, dramatic, musical or artistic work	Lifetime of the author plus sixty years after his death
Joint authorship	Sixty years after the last surviving author's death
Literary, dramatic, musical or artistic work (other than photograph), which is published anonymously or pseudo-anonymously	Sixty years from the year of first publication
Photograph	Sixty years from the date of publication
Sound recording and cinematograph films	Sixty years from the year of publication
Government work	Sixty years from the year of publication
Work done under public undertaking	Sixty years from the date of publication
International Organization	Sixty years from the date of first publication
Performers' Right	Twenty five years from the year in which the performance is made
Broadcasting Right	Twenty-five years from the year of first broadcast made

8. Rights

8.1 Copyright Protection is Bundle of Rights

Copyright is the right given to creators in their work, where the creator holds the exclusive rights to use or license others to use the work on agreement. It covers a broad range of classes and different rights subsist in different classes, thereby referred as a bundle of rights. The rights vary according to the class of work. It includes moral rights, economic rights, rights of reproduction, communication to the public, adaptation right, right of performance, broadcast and recording and translation of the work.

8.2 Rights in the Case of a Literary Work

In the case of a literary work (except computer programme), copyright means the exclusive right to: i) reproduce the work; ii) issue copies of the work to the public; iii) perform the work in public; iv) communicate the work to the public; v) make cinematograph film or sound recording in respect of the work; vi) make any translation of the work; vii) make any adaptation of the work.

Adaptation means the preparation of a new work in the same or different form based upon an already existing work which includes a) Conversion of a dramatic work into a non dramatic work; b) Conversion of a literary or artistic work into a dramatic work; c) Re-arrangement of a literary or dramatic work; d) Depiction in a comic form or through pictures of a literary or dramatic work; e) Transcription of a musical work or any act involving re-arrangement or alteration of an existing work; f) The making of a cinematograph film of a literary or dramatic or musical work is also an adaptation.

Computer programmes are protected under the Copyright Act as literary works. In addition to all the rights applicable to a literary work as mentioned above, owner of the copyright in a computer programme enjoys the rights to sell or give on hire or offer for sale or hire, regardless of whether such a copy has been sold or given on hire on earlier occasion.

8.3 Rights in a Dramatic Work

It gives the exclusive right to: a) reproduce; b) communicate or perform the work in public; c) issue copies of the work to the public; d) include the work in any cinematograph film; e) make any adaptation of the work; f) make translation of the work.

8.4 Rights in an Artistic Work

In the case of an artistic work, copyright means the exclusive right to: a) reproduce; b) communicate the work to the public; c) issue copies of the work to the public; iv) include the work in any cinematograph film; v) make any adaptation of the work.

8.5 Rights in a Musical Work

Copyright means the exclusive right to: a) reproduce the work; b) issue copies of the work to the public; c) perform the work in public; d) communicate the work to the public; e) make cinematograph film or sound recording in respect of the work; f) translation of the work; g) any adaptation of the work.

8.6 Rights in a Cinematograph Film

In the case of a cinematograph film, copyright means the exclusive right to: a) make a copy of the film including a photograph of any image forming part thereof; b) sell or give on hire or offer for sale or hire a copy of the film; c) communicate the cinematograph film to the public

8.7   Rights in a Sound Recording

The rights are: a) To make any other sound recording embodying it; b) To sell or give on hire, or offer for sale or hire, any copy of the sound recording; c) To communicate the sound recording to the public.

Communication to the public generally means that work is seen or heard or otherwise enjoyed by the public directly or by any means of display or diffusion. But, it is not necessary that any member of the public actually sees, hears or otherwise enjoys the work so made available. For example, a cable operator may transmit a cinematograph film, which no member of the public may see. It is still a communication to the public because the work in question is accessible to the public.

Things to remember – Copyright does not exist on:

There is no copyright over news. However, there is copyright over the way in which a news item is reported.

Copyright does not exist on yoga. [See Box 1 – a case study].

Box 1
In a significant ruling, Delhi high court in 2014 has held that exclusive rights over yoga and pranic exercises cannot be claimed under the Copyright Act. The high court rejected the plea of a Philippines-based Institute for Inner Studies seeking to restrain some persons from teaching the 'asanas' (postures) claimed to be developed by the founder of the institute. Justice Manmohan Singh delivered his verdict on a petition filed by the Institute, which was established by Late Samson Lim Choachuy, Master Choa Kok Sui, on April 27, 1987 and has trusts in various cities in India. The institute had moved court seeking ban against Charlotte Anderson and others.

Law permits any use of a work without permission of the owner of the copyright subject to certain conditions like a fair deal for research, study, criticism, review and news reporting of current events, judicial proceedings as well as use of works in library and schools and in the legislatures. Performance by an amateur club or society if the performance is given to a non-paying audience and the making of sound recordings of literary, dramatic or musical works under certain conditions.

The owner of the copyright in an existing work or prospective owner of the copyright in future work may assign his rights to any person, either partially or wholly, with imposed conditions of the author. However, in respect of future work, assignment takes effect after the work comes into being. A license can be granted in respect to a future work, but the licence will take into effect only when the work comes into existence.

Copyrights of works of the foreign countries mentioned in the International Copyright Order are protected in India, as if such works are Indian works. Also, copyright of nationals of countries who are members of the Berne Convention for the Protection of Literary and Artistic Works, Universal Copyright Convention and the TRIPs Agreement are protected in India through the International Copyright Order. Copyright as provided by the Indian Copyright Act is valid only within the borders of the country. Likewise, to secure protection to Indian works in foreign countries, India has become a member of international conventions on copyright and neighbouring (related) rights: i) Berne Convention for the Protection of Literary and Artistic works; ii) Universal Copyright Convention; iii) Convention for the Protection of Producers of Phonograms against Unauthorised Duplication of their Phonograms; iv) Multilateral Convention for the Avoidance of Double Taxation of Copyright Royalties: and v) TRIPs Agreement.

8.8 Moral Rights in Copyright Protection

The following are the moral rights:

1. The right to decide whether or not to publish the work.
2. The right to claim authorship of a published or exhibited work (the right of paternity).
3. The right to prevent alterations and other actions such as distortion, mutilation, modification or other acts which may damage the author's honour or reputation (the right of integrity).
4. The moral rights are independent of the author's copyright and remains with him even after assignment of the copyright. These rights remain with the author even after the transfer of copyright and the protection lasts during the whole of the copyright term. If the author desires to transfer these rights also he can do so specifically in writing in the deed of assignment. Moral rights are available to the authors even after the economic rights are assigned. Economic rights provide exclusive rights to the author to reproduce or adapt or communicate his work to the public. They enable an author to make pecuniary gains by controlling those rights, i.e. he/she can negotiate financial returns for licensing the use or for assignment of any of those rights.

9. Performers Rights

It is a special right provided to the performer engaged in any performance. The right shall subsist for twenty five years from the date in which the performance has been made. A performer includes an actor, singer, musician, dancer, acrobat, juggler, snake charmer, a person delivering a lecture, etc. Performances also include any visual or acoustic presentation made live by a performer. A performer has the right to make a sound recording or visual recording of the performance; right to reproduce the sound recording or visual recording of the performance; right to broadcast the performance; and right to communicate the performance to the public otherwise than by broadcast. It must be noted that once a performer has consented for incorporation of his performance in a cinematograph film, he/she shall have no more performer's rights to that performance.

10. Broadcasting Rights

Broadcasting is defined as communication of work to public by any means of wireless diffusion or by wire or any media for communication of audio, video, text, signs, sounds or visual images. The government or other broadcasting authority will be the owner of the broadcast reproduction right, which will subsist for 25 years from the year of broadcast. Certain broadcasting reproduction rights subsist in programmes broadcasted by the Broadcasting Authority either in public interest or for copyright materials with due consent of its owners. These rights are:

1. The right to re-broadcast the programme or any substantial part of it.

2. Causing the programme or any substantial part of it to be heard or seen by public on payment of any charges.

3. To make any sound or visual recording of the broadcast in question or any substantial part of it.

4. Right to sell or hire to the public, or offer for such sale or hire, any sound recording or visual recording of the broadcast.

5. Right to make any reproduction of such sound recording or visual recording where such initial recording was done without license or, where it was licensed, for any purpose not envisaged by such license.

11. Transferability and Assignment of Copyright

Transferability is necessary for proper exploitation of the rights. In copyright the transfer takes place through assignments. An assignment is in essence a transfer of ownership even if it is partial. This assignment is required to be in writing. An assignee can further assign the same to other person.

Assignment of copyright: Sec. 18 of the Copyright Act, 1957 deals with assignment of copyright. The owner of the copyright in an existing work or the prospective owner of the copyright in a future work may assign to any person the copyright either wholly or partially and either generally or subject to limitations and either for the whole term of the copyright or any part thereof. The mode of assignment should be in the following manner:

- Assignment should be given in writing and signed by the assignor or by his duly authorized agent.

- The assignment should indentify the work and specify the rights assigned and the duration and territorial extent of such assignment.

- The assignment should also specify the amount of royalty payable, if any, to the author or his legal heirs during the currency of the assignment and the assignment may be subject to revision, extension or termination on terms mutually agreed upon by the parties.

- Where the assignee does not exercise the rights assigned to him within a period of one year from the date of assignment, the assignment in respect of such rights will be deemed to have lapsed after the expiry of the said period unless otherwise specified in the assignment.

The period of assignment will be deemed to be 5 years from the date of assignment unless specifically mentioned. If the territorial extent of assignment of the rights is not specified, it will be presumed to extend within India.

If any dispute arises with respect to the assignment of any copyright the Copyright Board may, on receipt of a complaint from the aggrieved party and after holding such inquiry as it considers necessary, pass such order as it may deem fit including an order for the recovery of any royalty payable. In case the assignor is also the author, provided further that no order of revocation of assignment, be made within a period of five years from the date of such assignment.

12. Licensing

A copyright owner can permit another person to do certain acts without assignment, which is called licensing. In case of licensing the ownership does not get transferred. It is a permission to do something beyond which the license would be an infringement. Copyright can also be transferred by testamentary disposition or by the operation of law as in case of other properties.

Provisions with regard to licenses are detailed in Chapter VI of the Copyright Act, 1957. Copyright License is granted by the owner of the copyright in any existing work or the prospective owner of the copyright in any future work in writing signed by him or by his duly authorized agent. In the case of a license relating to copyright in any future work, the license will take effect only when the work comes into existence. Where a person to whom a license relating to copyright in any future work is granted dies before the work comes into existence, his legal representatives, in the absence of any provision to the contrary in the license, will be entitled to the benefit of the license.

Any person can apply for the grant of copyright license in Form II to produce and publish translation of a literary or dramatic work in any language in general use in India after a period of three years from the publication of such work, if such translation is required for the purpose of teaching, scholarship or research.

The Copyright Board after holding an enquiry may direct the registrar to grant copyright license to the person to publish the work or translation thereof in the language mentioned in the application. The applicant should deposit the amount of royalty as specified by the Copyright Board in the account of the original owner of the work.

The copyright license will be terminated at any time after the granting of a license to produce and publish the translation of a work in any language, if the owner of the copyright in the work or any person authorized by him publishes a translation of such work in the same language and which is substantially the same in content at a price reasonably related to the price normally charged in India for the translation of works of the same standard on the same or similar subject.

No termination will take effect until after the expiry of a period of three months from the date of service of a notice in Form IIB on the person holding such license by the owner of the right of translation intimating the publication of the translation as aforesaid.

If the owner of a copyright at any time during the term of copyright has refused to allow the re-publication of that work or has refused to allow the performance in public (withheld) or has refused to allow communication to the public by broadcast, a complaint can be made to the Copyright Board. If satisfied the Board may direct the Registrar of the Copyright to issue a compulsory license to republish, perform, or broadcast the work to the public after the payment of compensation to the owner of copyright, provided that the work is an Indian work. In case two or more persons have made a complaint for the same work, then license will be issued to the person who would best serve the interest of the general public.

The International Treaties such as Berne Convention and Universal Copyright Convention provides a special provision for non-voluntary license also for the benefit of developing countries. These licenses are confined to translation and reproduction rights and are temporary since they are permissible as long as the country concerned ranks as a developing country.

For unpublished work, a compulsory license can be issued. If the author is dead, unknown or cannot be traced, in such case, any person can apply to the Copyright Board for a license to publish or translate that work. The conditions are:

1. The applicant will have to publish his proposal in one issue of a daily newspaper in English language having a wide circulation.

2. For translation issues it has to be published in daily issue of that language.

After the grant of a license, the applicant is directed to deposit the amount of royalty in the public account of India so as to enable the owner of the copyright, his heirs, or legal representatives who may claim such royalty at any time.

With regard to license for translation of published work, any person can apply to the Copyright Board. In case of Indian work, application can be made after a period of seven years from the first publication of work. For works other than Indian work an application can be made after a period of three years from the date of first publication if required for teaching, scholarship or research. If such translation is not in a language in general use by a developed country, an application can be made within a period of one year from the year of its first publication. Applicant shall pay a royalty to the owner of the copyright in respect to the number of copies sold of the translation. Not to be exported outside India and such translation should also carry a notice for distribution in India only. Such translations are liable for export if:

1. Translation is in a language other than English, French and Spanish.

2. Such copies are to be used for the purpose of teaching, scholarship or research and not for commercial purpose.

3. Proper permission has been obtained from the Government of India.

13. Fair Dealing and Fair Use Provision

Fair dealing means how much of the reproduction is fair which does not infringe the rights of a holder. It depends on value of matter into consideration, its purpose, likelihood of competition between the two works and whether the work is published or not. If extracts taken are used as a basis for comment, review, research, private study, criticism or review reporting current events in the newspaper or criticism then it is fair dealing but if used for rival purposes then it is unfair.

Copyright law permits certain uses of copyright protected work without any specific authorization of the owner and at no cost. These are referred as fair use provisions. Berne Convention prescribes the following permissible exceptions:

i. It should not conflict with normal exploitation of the work.

ii. It should not unreasonably prejudice the legitimate interest of the author.

iii. Such permissible reproduction should be only in certain special cases.

iv. Using certain copyright material for reporting of current events.

v. Legitimate use of new copyrights works for the educational, scientific and cultural advancement of the society.

Regarding publication of foreign books in India, if relevant period of expiry is over from the date of first publication of an edition of a literary, scientific or artistic work, provided:

i. Copies of such edition are not available in India.

ii. Copies have not been put for sale in India for a period of six months to the general public.

iii. Not sold in connection with systematic instruction activities at a price reasonable related to than charged in India.

In such case a license can be issued to the applicant provided a royalty is paid to the owner of the copyright and with the condition "Not for sale outside India".

The relevant period of expiration is:

i. Seven years from the date of first publication for reproduction and publication of any work related to fiction, poetry, drama, or music art.

ii. Three years from the date of first publication of the work in case of natural science, physical science, mathematics or technology.

iii. Five years from the date of first publication of work in any other case.

However, no foreign book can be copied for use without the consent or prior information of the author(s), even if it is not available for sale in India. But for a library under the direction of librarian only three copies of that book (also includes, pamphlet, sheet of music, map, chart or plan) can be made for the use in the library.

14. Infringement of Copyright

Copyright in a work can be infringed when in general work is commercially exploited in any form by a person without authority. In general following are some of the situations when copyright is infringed by a person:

i. Making infringing copies for sale or hire or selling or letting them for hire;

ii. reproduces and/or publishes the work in a material form;

iii. makes adaptations and translations of the work without due authority or permission of the copyright owner

iv. makes use of a copyright material without license from the owner of the copyright, or the Registrar of Copyright (in certain situations) or in contravention of the conditions of a license

v. permitting any place for the performance of works in public where such performance constitutes infringement of copyright;

iv. distributing infringing copies for the purpose of trade or to such an extent so as to affect prejudicially the interest of the owner of copyright;

vii. public exhibition of infringing copies by way of trade; and

viii. importation of infringing copies into India.

However, the following acts do not constitute infringement of copyright:

i. Fair dealing for research or private study, criticism or review.

ii. Reproduction for use in judicial proceedings and by members of legislature.

iii. Publication of short passages, restricted reproduction or performance for educational purposes.

iv. Making of records under license from Copyright Board on payment of royalty.

v. Playing of records or performance by a club or a society for the benefit of members of religious institutions. (However, telecast of certain programmes e.g. telecast of matches etc. in hotels/restaurants etc. where there is an assembly of more than 50 persons then a license has to taken from the organization which own the telecast rights).

vi. Reproduction of an article on current economic, political, social or religious matters in newspapers, magazines, etc.

vii. Reproduction of few copies for use in libraries or for research or private study.

viii. Matters published in official Gazettes including Act of Parliament or its translation.

ix. Making of a drawing, engraving, or photograph of an architectural work of art, or sculpture kept in a public place.

x. Use of artistic work in a cinematograph film.

xi. Use of an artistic work (author not the owner of copyright) by the author of any mould, cast, sketch, model, etc made by him for the work.

xii. Making of an object in 3-D of an artistic work in 2-D subject to conditions.

xiii. Reconstruction of a building in accordance with architectural drawings.

15. Penalties for Infringement

A copyright owner can take legal action against any person who infringes the copyright in the work. The copyright owner is entitled to remedies by way of injunctions, damages and accounts. A suit or other civil proceedings relating to infringement of copyright should be instituted in the District Court or High Court within whose jurisdiction the plaintiff resides or carry on business. No court inferior to that of a Metropolitan Magistrate or a Judicial Magistrate of the first class shall try any offence under the Copyright Act. The period of limitation for filing the suit is three years from the date of infringement in case of civil suit. Any person who knowingly infringes or abets the infringement of the copyright in any work commits a criminal offence under section 63 of the Copyright Act, 1957. Any police officer, not below the rank of a sub inspector, if satisfied that an offence in respect of the infringement of copyright in any work has been/is being/or is likely to be committed, seize without warrant all copies of the work and all plates used for the purpose of making infringing copies of the work, wherever found, and all copies and plates so seized shall, as soon as practicable, be produced before a magistrate.

Punishment for any person who knowingly infringes or abets the infringement of copyright in a work or any other related rights will be punished for a term of not less than six months which may extend to three years and with a fine of rupees not less than fifty thousand, extendable up to rupees 1 lakh. For second and subsequent conviction, the form of punishment will be not less than one year and up to three years and a fine of minimum of rupees one lakh extendable up to 3 lakh rupees.

Any person who uses an infringing copy of a computer programme shall be punished with an imprisonment not less than seven days extendable up to three years and a fine of not less than fifty thousand rupees, which may extend to two lakh rupees.

If a company has committed any offence, every person who at the time of offence was in charge and was responsible to company for the conduct of the business of the company, shall be considered guilty of such offence and is liable to be punished accordingly. If any person in such company proves that the offence was created without his knowledge or had tried to prevent the operation of such offence may be omitted from the offence.

16. Case Studies on Infringement of Copyrights

Zee Told to Pay Copyright Damages: Delhi high Court in 2009 has ordered Zee Telefilms Ltd. to pay past dues and the license fee amount to T-Series in a case of alleged copyright infringement over the former's music show Sa Re Ga Ma Pa. T-Series contended that Zee had been using its content without a valid license. In April 2009 Zee had entered was over on March 31 but there was no renewal for the period commencing from April. Zee was not willing to pay the license fee or the outstanding amount of Rs 69 lakh and continued to use the copyrighted content of T-Series into a licensing arrangement with T-Series for using its copyrighted content in its programmes. The license period even after the expiry of the licence period.

Rakesh Roshan settles copyright dispute on Krazzy 4: Rakesh Roshan's movie, Krazzy 4, was restrained by High court in 2008 with regard to the songs 'Break-Free' the title track and their remixes on a complaint filed by Ram Sampath (advertisement-jingle composer). It is believed that Rakesh Roshan had taken permission from Ericsson for using the tune in its movie but came to know later when the case was filed by Sampath that Ericsson was not the legal owner of the tune. Finally Rakesh Roshan reached a settlement with Ram Sampath for allowing the use of the tune in its movie with a settlement amount of Rs 2 crores as reported in the media.

Disney loses Winnie the Pooh: US media giant Walt Disney Co has lost a court battle in a long-running fight over the copyright of the Winnie the Pooh character. A US Federal judge in California in the year 2007 granted Stephen Slesinger Inc, which claims the rights to Winnie the Pooh, a 'summary judgment' that effectively ends Disney's efforts to take back the copyright. The heirs of Stephen Slesinger, who bought the US rights from 'Pooh' author AA Milne in 1930 and began licensing them to Disney in 1961, claim the powerful firm has cheated them out of hundreds of millions of dollars in royalties.

Zandu Balm – Dabangg case: Zandu Pharmaceuticals, makers of Zandu Balm have served a legal notice on Arbaaz Khan producer of the film 'Dabangg' for copyright infringement of the product 'Zandu Balm'. Dabangg features a song called *Munni badnam* which uses the word *zandu balm*. Report suggests that Zandu Pharmaceuticals is claiming that it's the sole copyright holder of "not only the product, but also the name Zandu Balm" and asked the producers to withdraw the song or delete the name of the product from it, failing which it might take action against them. Surprisingly, no issue of trademark infringement was brought in since the legal notice says so. In certain strange and old cases of the past, copyright protection has been however granted to a string of words. However, this seems to be a case of nominative use of trademark where the use of *zandu balm* does not suggest that it is from the movie *Dabangg* so as to amount to any endorsement. Zandu's lawyer Ashok M Saraogi sent legal notice to Arbaaz Khan, his partner and wife Malaika Arora (Munni in the movie), Shri Ashtavinayak Cinemas Ltd, and director Abhinav Kashyap. The notice adds, *"In the song, the name Zandu balm has been used continuously at various places. By using the brand name in the song, you have not only violated the copyright of my clients, but you have also made an attempt to defame the reputation of my clients and the product manufactured by them."* Very surprisingly, the objection comes in a stage when Dabangg has scored big at the box office and the music album is the highest selling currently. Reports have also suggested that the sales of Zandu Balm have actually shot up after the release of the movie in 2010. However, Zandu claims that they also launched a new ad-campaign around that time and thus the sales figure could be a consequence of that. Recently, Lalit Pandit, who is the composer of the song *Munni badnam* credited the word *Zandu* to Madhur Bhandarkar of *Fashion* fame, who apparently uses the word in his everyday language. Reports indicate that out of court settlement was there.

Scrabble-maker sues Scrabulous.com: US toymaker Hasbro of the popular word-game, Scrabble have sent legal notices to two Kolkata-based brothers Jayant and Rajat Agarwalla, who have expanded Scrabble into an online version called Scrabulous and put it on a trendy social networking site in June 2007 for copyright violation. Within months, Scrabulous version became wildly popular, with some 2.3 million people using it everyday. The online versions allow players to engage several games simultaneously across countries and continents; the games, with expanded rules, can be played over days, instead of the standard two hours. That attracted the attention of Scrabble-maker Hasbro, which has sold the online rights to Electronic Arts (EA). Hasbro had filed a case of infringement of the Digital Millennium Copyright Act in U.S. District court in New York, naming the brothers, Jayant and Rajat Agarwalla, as the defendants along with their web design and technology company R.J. Software. Hasbro owns Scrabble's North American and Canadian rights. Mattel, which owns the Scrabble brand outside the U.S. and Canada, is already pursuing a lawsuit against Scrabulous.com in Indian courts over trademark and copyright violation issues. There are several versions of Scrabble online but the Agarwalla knock-off has caught on like wildfire, riding on the frenzy associated with the social networking site. The brothers reportedly wrote to Hasbro seeking permission to put their version online but did not hear from the toy-maker. Meanwhile, Scrabulous.com has stopped its fans in the U.S. and Canada from accessing the free add-on Scrabulous application on Facebook till further notice.

17. Copyright in Design

The exclusive right conferred on a design is termed as 'Copyright in Design'. This is different from the copyright in artistic and literary work. There may be certain designs, which can qualify for registration, both under Designs Act, 2000 and Copyright Act, 1957. The Designs Act, 2000 covers the industrial design and product design. If a design has been registered under the Designs Act, the Copyright Act cannot protect it even though it is an original artistic work. If a design qualifies for registration under Designs Act, but has not been registered under Designs Act, the exclusive right will subsist under the Copyright Act. Copyright Act will cease to exist when the article to which the design has been applied and reproduced more than fifty times by an industrial process by the owner of the copyright under Copyright Act.

There is an overlap between the two Acts. The Copyright Act provides that if a design is registered under the Designs Act, then it does not get protection under the Copyright Act. Those, which are not registrable under designs, will be entitled for substantial protection under the Copyright Act. Artistic works can be reproduced in any material form, which includes duplication in a three-dimensional or two-dimensional form. As long as designs falls under the definition of "artistic works" within the meaning in the Copyright Act it is afforded vast protection. It should be noted that obtaining a copyright is easier than obtaining design protection. Added to this, greater rights and remedies are provided under the Copyright Act which have made the Designs Act an unfavourable option.

18. Copyright and Computer Software

Computer programmes can be protected under the head "literary work" of the Copyright Act, 1957. According to the Indian Copyright Act, 1957 a "Computer" includes any electronic or similar device having information processing capabilities. "Computer programme" means a set of instructions expressed in words, codes, and schemes or in any other form, including a machine-readable medium, capable of causing a computer to perform a particular task or achieve a particular result.

The Indian Copyright (amendment) Act, 1994 which was brought into effect on 10 May, 1995 included provisions for "protection of computer software" and "computer generated works" (by considering them as literary works). Computer software as "literary work" included computer programmes and computer databases. For the first time in India, the Copyright Law clearly explained the rights of a copyright holder, position on rentals of software and the rights of the user to make backup copies. Since most software is easy to duplicate, and the copy is usually as good as original, the Copyright Act amendment was needed.

Some of the key aspects of the law are:

- According to section 14 of this Act, it is illegal to sell or give on commercial rental or offer for sale or for commercial rental any copy of the computer programme. The violator can be tried under both civil and criminal law.

- A civil and criminal action may be instituted for injunction, actual damages (including violator's profits) or statutory damages per infringement, etc.

- Heavy punishment and fines for infringement of software copyright.

- Section 63 B stipulates a minimum jail term of 7 days, which can be extended up to 3 years and with fine which shall not be less than fifty thousand rupees but which may extend to two lakh rupees.

Intellectual property protection of databases is one of the important issues for scientists, researchers and innovators. Databases are collections or compilation of records that are organized for easy access and retrieval. The growth of databases in electronic formats has increased the need for their legal protection. In most countries, databases qualify for IP protection through copyright and trademark legislation. They may also be protected *de facto* through a contract system between the database supplier and the user. Protection of digital IP can be achieved by applying legal, administrative as well as technological measures. The technological measures for the protection of digital IP could be implemented through (i) access control, (ii) control of certain users, (iii) integrity protection, (iv) usage metering, and (v) electronic copyright management.

With technological improvements, new methods have emerged through which products of creative activity can reach out to the public. The invention of computers and computer programmes has led to introduction of digital technology. This has brought a revolution in the concept of copyright protection. The traditional means of transmission of work such as books, video records, etc. are being overtaken by new interactive online systems which enable to access the databases through wireless or cables. Strengthening of the intellectual property system in the context of internationalization of trade, of Internet economies, of the multimedia age, and of information technology in the current century, is a constant necessity.

Software sector is one of the fastest growing areas of the information technology market. With the availability of the modern copying equipment, and with pirates having equal access to the new technologies, book publishers, music publishers, record manufacturers, film producers as well as software producers, are faced with pirate competition on a very high scale. Built-in information signals in the digitized versions of books, music, films, etc. could be read through electronic devices provided in the equipment made available to the public, and could render it possible to control the extent of utilization of work of a copyright owner. A legal framework for protection of software under national intellectual property laws is therefore, essential and so is its active and effective enforcement to prevent copying without compensation, or prior authorization, except for certain free uses of work.

Most computer software contains trade secrets. The programmer prepares the software in source code, but the software distributed to customers is generally mere executable code. The internal format and structure of the executable code is relatively uninterruptible by humans, which to some extent protects trade secrets contained in the source code. In India computer software/ databases can be protected under the Copyright Act where the owners of databases take precaution that the licensed user may use it for the value paid and do not get complete database to exploit it by making minor modifications. In advanced countries, it is possible for the author of computer software simultaneously to assert trade secrets in the source code, and to assert copyright rights in the source code (in executable code).

It is a new movement in the area of computer software, particularly with respect to the operating system. It seeks to freely share intellectual work with other like-minded programmers. The software designer provides a license to manufacturers and vendors who agree to its licensing terms. The idea behind open source software is that any programmer, who has received a copy of the software and has agreed in a licensing contract to its conditions, can adapt, change, modify, reproduce and disseminate the operating system. It does not mean that the software is necessarily free, publicly owned, or without significant limitations on use. The mindset behind this is a sort of collective development; more programmers focusing on the program will bring swifter upgrading, a quicker fix for the problems and bugs, and make a better program; and nobody will own the addition, upgradation, or modification. The difficulties which lies in the open source software system is that there is no central authority to confirm or reject modifications made by a number of programmers and there is a limitation to technical support and warranties to modified versions.

Relying on the rights provided in the WCT and WPPT, the computer software industries has sought out methods to prevent illegal offer of their product to the public. The industry has used the courts and law enforcement agencies, and also has created special programmes like web crawler software (developed by Microsoft, USA) to seek out websites, which illegally offer computer software. They use the technology called steganography, also known as watermarking. Through this method, information can be embedded in the copyrighted work such as music, films, software and books in digital or analog form, allowing the copyright owners to determine whether the work has been illegally copied. However, because new piracy operations are borne or resurfaced daily, this is a never-ending fight, which requires huge resources and efforts. Digital watermark is the process of modifying image data by inserting codes for carrying information. Watermarking of contents is carried out to ensure: (i) copyright protection, (ii) data authentication, and (iii) ownership identification.

There are four major software industries that have earned billions of dollars by licensing their computer software programmes. They are Microsoft, IBM Corporation, Computer Associates and Oracle Corporation. All the information resources utilized in the course of any organization's business is referred as an information asset and includes all the information, applications (software developed or purchased), and technology (hardware, system software and network). The Government of India has passed an Information Technology Act, 2000 to facilitate e-commerce. This Act provides legal recognition for transactions carried out by means of electronic data interchange and other means of electronic communication, which uses alternative to paper-based methods for communication and storage of information to facilitate electronic filing of documents with the Government agencies.

19. Administration of Copyright Law

The Copyright Office is under the immediate control of the Registrar of Copyrights who shall act under the superintendence and direction of the Central Government. The Registrar of Copyrights has the powers of a civil court when trying a suit under the Code of Civil Procedure in respect of the following matters, namely:

  a. summoning and enforcing the attendance of any person and examining him on oath;

  b. requiring the discovery and production of any document;

  c. receiving evidence on affidavit;

  d. issuing commissions for the examination of witnesses or documents;

  e. requisitioning any public record or copy thereof from any court or office;

  f. any other matters which may be prescribed.

19.1 Copyright Board

The Copyright Board in India, a quasi-judicial body, was constituted in September 1958. The jurisdiction of the Copyright Board extends to the whole of India. The Board is entrusted with the task of adjudication of disputes pertaining to copyright registration, assignment of copyright, grant of Licenses in respect of works withheld from public, unpublished Indian works, production and publication of translations and works for certain specified purposes. It also hears cases in other miscellaneous matters instituted before it under the Copyright Act, 1957. The meetings of the Board are held in five different zones of the country. This facilitates administration of justice to authors, creators and owners of intellectual property including IP attorney's near their place of location or occupation.

19.2 Powers of the Copyright Board

The Copyright Board consists of a Chairman and two or more, but not exceeding fourteen, other members for adjudicating certain kinds of copyright cases. The Chairman of the Board is of the level of a judge of a High Court. The Board has the power to:

i. hear appeals against the orders of the Registrar of Copyright;

ii. hear applications for rectification of entries in the Register of Copyrights;

iii. adjudicate upon disputes on assignment of copyright;

iv. grant compulsory licenses to publish or republish works (in certain circumstances);

v. grant compulsory license to produce and publish a translation of a literary or dramatic work in any language after a period of seven years from the first publication of the work;

vi. hear and decide disputes as to whether a work has been published or about the date of publication or about the term of copyright of a work in another country;

vii. fix rates of royalties in respect of sound recordings under the cover-version provision; and

vii. fix the resale share right in original copies of a painting, a sculpture or a drawing and of original manuscripts of a literary or dramatic or musical work

19.3 Copyright Enforcement Advisory Council (CEAC)

The Government has set up on November 6, 1991 a Copyright Enforcement Advisory Council (CEAC) to review the progress of enforcement of Copyright Act periodically and to advise the Government regarding measures for improving the enforcement of the Act.

19.4 Register of Copyrights

A register of copyright is kept at the Copyright Office for entering the names or title of works, names and addresses of authors, publishers and owners of copyright and other particulars of the work. The entry is made only after the payment of a prescribed fees and successful enquiry. This register of copyrights acts as a prima facie evidence of the particulars entered therein, since it is certified by the Registrar of Copyright and sealed with the seal of the Copyright Office. It can be a proof in all courts without any proof or production of the original. Every entry, correction of the entries and rectification ordered in the register of copyright are published in the official Gazette by the Registrar of Copyright. The register of the copyright is kept in six parts.

Part I   Literary Work

Part II   Musical Work

Part III   Artistic Work

Part IV   Cinematographic files

Part V   Sound recordings

Part VI   Computer programmes, tables and compilation including computer databases.

Selected reading

Adukia, R. 2012. Handbook on Intellectual Property Rights www.metastudio.org/Science/

Chawla, H.S. (2014). Copyright Protection. In: Redefining Libraries in Digital Era (Eds. Superna Sharma, Arundhati Kaushik, Chanda Arya and Hema Haldua), Astral International Pvt Ltd., Delhi, pp. 184-197.

Chawla, H.S. and Singh, A.K. 2005. Intellectual Property Rights. Vol II: Copyrights, Trade Marks, Trade Secrets and Geographical Indications. Pantnagar University Press, pp.75

http://www.wipo.int/treaties/en/

The Copyright Act, 1957 along with The Copyright Rules and International Copyright Order 1999, Universal Law Publishing Co., Delhi, 2011

Trade Marks

Trade mark is a mark represented graphically which is capable of identifying and distinguishing the goods or services of one person from those of others that may be a word, name, symbol, label, design or device, or any combination of colours, shape of goods, used or intended to be used in commerce. A trade mark is a brand name which is used for products. Trade Marks are distinctive symbols, signs, logos that help consumer to distinguish between competing goods or services. On the other hand a trade name is the name of an enterprise/business entity which individualizes the enterprise in consumer's mind. It is legally not linked to quality, but linked in consumer's mind to quality expectation. Simply put, a trade name is the official name under which a company does business.

Marks have been used to identify the source of goods for a long time. There is evidence that 4,000 years ago, craftsmen from China, India, and Persia used either their signatures or symbols to identify their products. Roman pottery-makers used more than 100 different marks to distinguish their work. These craftsmen are believed to have used marks for several purposes, including as an advertisement for the makers of the products, as proof that the products belonged to a particular merchant in the event of an ownership dispute, and as a guarantee of quality. In the middle ages, the use of marks eventually became associated with the development and growth of skilled trades, and hence the term "Trade Marks." Marks were used to show that members of a society were known to have experience in the trade. In modern times, trade marks have developed into identifiers of products from individual companies and are important business assets. As branding has become an important marketing concept, the legal protection afforded to trade marks has grown in importance as well. Today "trade mark" has become almost synonymous with "brand".

1.   International Treaties and Conventions for Trade Marks

1.1   *Madrid Agreement and Madrid Protocol*

Madrid Agreement concerning the International Registration of Marks came into force in 1989 and was meant for repression of false or deceptive indications of source on goods. The main features of the Madrid Agreement are as follows:

1. An applicant must be a national of a member country. A person having his domicile or a real and effective industrial or commercial interest in such a country is also eligible. It may be noted that this would be governed by the national laws of the country in question.

2. A mark to be registered in member states should be first registered at the national level in the country of origin of the applicant. The first registration is called 'basic registration'.

3. The country having given the basic registration can only transmit the request for international filling to the International Bureau of the World Intellectual Property Organization (WIPO) along with the list of the courtiers in which protection is being sought. There is no provision for directly filing a request under the agreement.

4. It is required that the country of origin has to be a member state. The role of the office of the country of origin is not only to send the application for international registration but also to certify that the mark, which is the subject of the international registration, is the same mark, which has been registered in the country of origin.

5. For each application, a fee has to be paid for each designated country and WIPO. The fee paid for the designated countries is called the 'complementary fee'.

6. The International Bureau notifies the international registration to the offices of the designated countries and publishes it in a monthly periodical called 'The WIPO Gazette of International Marks'.

7. If the basic registration is cancelled for some reasons, in the country of origin, during the first five years, the international registration automatically stands cancelled in all the designated countries. This also gives an advantage to a person to oppose the registration of a mark only in the country of origin and that person need not oppose it in all the designated countries. This possibility of challenging an international registration through a national registration is referred to as 'Central Attack' feature of the Agreement.

Madrid Protocol: It relates to the Madrid Agreement concerning the International Registration of Marks entered into force on December 1, 1995 and came into operation on April 1, 1996. It has the basic features of the Madrid Agreement. The protocol was formed to remove some of the features of the Madrid Agreement, which posed some obstacles to accession by several countries which are as mentioned below:

1. For an international registration, it is essential to first register a mark at the national level. The time required for obtaining a mark at the national level varies from country to country. Hence some parties do suffer.

2. Within one year, a designated member country has to examine and issue a notice of refusal by giving all the grounds for refusal. The period was considered short.

3. A uniform fee is paid for the designation of a member country. This was found to be inappropriate for countries with high level of national fees.

4. An international registration is linked to the basic registration during the initial five years and the former gets cancelled if latter is cancelled. The fact, that grounds under which a mark is cancelled in the country of origin need not necessarily exist in every other designated country, is overlooked.

5. The only working language of the Madrid Agreement is French.

Innovations introduced by the Madrid Protocol: Under the Madrid Protocol, a mark can be protected in many jurisdictions by filing an application for international registration. Such application is presented to the International Bureau of the WIPO at Geneva, through the office of origin i.e. the trademark office of the applicant. Where the application complies with the applicable requirements, the mark is recorded in the International Register and published in the WIPO Gazette of International Marks. The International Bureau then notifies each Contracting Party in which protection has been requested whether in the international application or subsequently. Each designated Contracting Party has the right to refuse protection of mark by so notifying to the International Bureau within the time limits specified in the Madrid Protocol. Unless such a refusal is notified to the International Bureau within the applicable time limit, the protection of the mark in each designated Contracting Party is the same as if it had been registered by the Office of that Contracting Party. An international registration remains dependent on the mark registered or applied for in the Office of origin, for a period of five years from the date of its registration. If, and to the extent that, the basic registration ceases to have effect within this five-year period, the international registration is no longer protected. An international registration subsists for the period of 10 years from the date of its registration and it may be renewed further by paying renewal fee before the expiry of every 10 years. All changes subsequent to the international registration, such as a change in name and/or address of the holder, a (total or partial) change in ownership of the holder or a limitation of the list of goods and services in respect of all or some of the designated Contracting Parties, may be recorded and have effect by means of a single procedure with the International Bureau and the payment of one fee.

After accession to the Madrid Protocol, the Trade Marks Registry (TMR) office of India will have two fold responsibilities:

i) As an office of origin it will receive International Applications, verify such applications are in conformity with the provisions of the Madrid Protocol and if the International Applications are proper the office will certify and transmit such applications to the International Bureau of WIPO. If any irregularity is found by the International Bureau of WIPO in any International Application transmitted from the TMR, it notifies such irregularities to the applicant (or his representative) as well as to the TMR.

ii) As an office of the designated contracting party the Indian TMR shall be notified about the international registrations in which India has been designated, this office shall record the particulars of such international registrations, examine it in accordance with the provisions of the Trade Marks Act & Rules. In case an objection is found during the examination of such international registration or an opposition is received after publication, this office shall communicate a provisional refusal (on the basis of examination or on the basis of opposition, as the case may be) to the International Bureau of WIPO.

With effect from 8th July 2013, the trademark filing in India is done according to the Madrid System of international trade mark registration. Thus, an application for the international registration of trademark can be filed under the Madrid Protocol.

1.2 Paris Convention for the Protection of Industrial Property (1883)

1. It provides for the right of priority in trade marks.

2. Domestic laws of the concerned country will govern the conditions for the filing and registration of marks.

3. If a mark has been duly registered in the country of origin, on request it must be accepted for filing and protected in its original form in other countries also.

4. It provides protection to collective marks.

5. It provides a provision for protection of trade names without the obligation of filing or registration.

6. It provides a provision for taking measures against direct or indirect use of a false indication of the source of the goods or the identity of the producer, manufacturer, or trader.

1.3 Nice Agreement concerning the International Classification of Goods and Services for the purpose of the Registration of the Marks (1957)

The International Classification consists of a list of classes (34 for goods and 8 for services) and an alphabetical list of the goods and services (11,000 items which are amended time to time). India also follows the same classification.

1.4 Vienna Agreement establishing an International Classification of the Figurative Elements of Marks (1973)

It establishes a classification for marks that consists of or contains figurative elements. The classification consists of 29 categories, 144 divisions and 1,796 sections in which the figurative elements of marks are classified.

1.5 TRIPs (1994)

It requires signatories to register service marks as well as trade marks. It provides protection for internationally well-known marks. It prohibits mandatory linking of trademarks and compulsory licensing of trade marks.

1.6 Trade Mark Law Treaty (1994)

Its aim was to make national and regional trade mark registration system more user friendly through simplification and harmonization of procedures and removing pitfalls present before the trade mark registry i.e. application for registration, changes after registration and renewal.

1.7 Singapore Treaty on the Law of Trademarks (2006)

The objective of the Singapore Treaty is to create a modern and dynamic international framework for the harmonization of administrative trademark registration procedures. Building on the Trademark Law Treaty of 1994 (TLT), the Singapore Treaty has a wider scope of application. The Singapore Treaty was concluded in 2006 and entered into force in 2009. The Singapore Treaty is applicable to all types of marks registrable under the law of a given Contracting Party. Most significantly, it is the first international instrument dealing with trademark law to explicitly recognize non-traditional marks. The Treaty is applicable to all types of marks, including non-traditional visible marks, such as holograms, three-dimensional marks, color, position and movement marks, as well as non-visible marks such as sound, olfactory or taste and feel marks.

1.8 Nairobi Treaty on the Protection of the Olympic Symbol (1981)

All States party to the Nairobi Treaty are under the obligation to protect the Olympic symbol – five interlaced rings – against use for commercial purposes (in advertisements, on goods, as a mark, etc.) without the authorization of the International Olympic Committee. An important effect of the Treaty is that, if the International Olympic Committee grants authorization to use the Olympic symbol in a State party to the Treaty, the National Olympic Committee of that State is entitled to a part in any revenue the International Olympic Committee obtains for granting the said authorization. The Treaty does not provide for the institution of a Union, governing body or budget.

2.   Trade Mark Law in India

Trade Mark law was enacted in India as Trade and Merchandise Marks Act, 1958 and now it has been governed by Trade Marks Act, 1999 and the Trade Marks Rules, 2002 amendment 2012.

The newly enacted Trade Marks Act, 1999 has some features not present in the Trade and Merchandise Marks Act 1958 and these are:

1. Registration of service marks, collective marks and certification trade marks.
2. Increasing the period of registration and renewal from 7 to 10 years.
3. Allowing filing of single application for registration in more than one class.
4. Enhanced punishment for offences related to trade marks.
5. Exhaustive definitions for terms frequently used.
6. Simplified procedure for registration of registered users and enlarged scope of permitted use.
7. Constitution of an Appellate Board for speedy disposal of appeals and rectification applications, which at present lie before High Court.

3.   Kinds of Trade Marks

- Any name (including personal or surname of the applicant or predecessor in business or the signature of the person), which is not unusual for trade to adopt as a mark (e.g. *Godrej*, Tata).
- An invented word or any arbitrary dictionary word or words, not being directly descriptive of the character or quality of the goods/service (e.g. *Rin, Surf*).
- Letters or numerals or any combination thereof e.g. Tata, Bata, Amul.
- The right to proprietorship of a trade mark may be acquired by either registration under the Act or by use in relation to particular goods or service.
- A trade mark which has acquired distinctiveness by use over a prolonged period of time (e.g. Xerox for photocopying, *Surf* as detergent powder).

- Devices, including fancy devices or symbols
- Monograms
- Combination of colors or even a single color in combination with a word or device (e.g. capsule colours)
- Shape of goods or their packaging
- Marks constituting a 3- dimensional sign.
- Sound marks when represented in conventional notation or described in words by being graphically represented. Yodeling sound has been registered by Yahoo. Now Aktiengesellschaft, a German Company (Allianz) has secured registration of a sound mark for Allianz, represented by a well know Indian firm in 2009. Over the years there has been a shift from conventional trade marks to the non-conventional such as smell, taste, sound, song tag line, etc. [See Box 1] as long as they are capable of being represented graphically, in accordance with the provisions of the Indian Trade Marks Act, 1999.

Box 1

"Kolaveri Di..." to be India's Trade Mark

Sony Music Entertainment India in 2012 recorded the super duper hit "Why this Kolaveri di..." may soon become the first song in India to have its first line trademarked. Recently, with an obvious intention of securing exclusive right to use the song tag line "Why this Kolaveri di..." filed for trade mark registration. The song tag line may just be the first ever song to be trademarked. The trade mark registration has been applied for in classes 9 and 41 which will allow Sony to launch products such as compact disks, cassettes and SD cards as well as film and non-film entertainment content and talent discovery programmes. "It is a smart move, a first of its kind in the Indian music industry,". "Kolaveri di..." song, written and sung by actor Dhanush for Tamil film '3', is now a case study on how viral marketing can create a cult following as it was watched by millions from all over the world within days of its digital release by Sony Music in the second week of November.

---

4. Essential Elements of Trademarks

A trade mark is considered good when it is distinctive. Distinctiveness may be inherent or acquired. A trade mark is said to be inherent when the mark or getup is distinct in itself from everything else and no one can justifiably claim the right to use it. Example: The mark in the shape of an invented word like 'Rin'. Acquired distinctiveness is when the trade marks acquire distinctiveness through use like Hawkins, *Surf*, Kodak, Yashica, etc. These are also distinctive due to inherent quality of being invented words. The ideal features of trade marks are:

1. It is preferable to use invented words (e.g. *Rin, Surf*).

2. It should be easy to pronounce and remember (e.g. SONY, *Zen*, etc.).

3. It should be capable of being described by a single word (e.g. Tata, *Bata*).

4. It must be easy to spell correctly and write legibly.

5. It should not be descriptive but may be suggestive of the quality of goods. Example A-1 generally suggests good/superior quality. Ex: Avon (A-1).

6. It should be short, e.g. *Rin, Zen*, etc.

7. It should appeal to the eyes as well as the ears, e.g. Titan, Hercules.

8. It should satisfy the requirements of registration.

9. It should not belong to the class of marks prohibited for registration. A mark contrary to law for time being in force or mark prohibited under the emblems and names such as ASC, RAW, AIR, WWF.

There are a large number of companies whose brand names have become the most valuable assets of a company, often exceeding the value of their physical assets. Unregistered trademark used for promotion of branded goods is designated by TM symbol while registered trademark is designated as ® Similarly unregistered service mark is written as SM. A few of them are McDonald™, AT&T™, Dolby™, Nike™, Adidas™, Kodak™, etc.

5. Functions of Trade Marks

It performs the following functions:

1. It identifies the goods/services and its origin. Example: 'Brooke Bond' identifies tea originating from the tea company manufacturing and marketing it under that mark.

2. It guarantees its unchanged quality. Example: The quality of tea sold in the packs marked Brooke Bond tea would be similar but different from tea labeled with mark Taj Mahal.

3. The trade mark advertises the product or goods. Example: 'SONY' is associated with electronic items, which tells us about the quality of particular class of goods. It thus advertises the product while distinguishing it from products of SONY's competitors.

4. It creates an image of the product in the minds of consumers and prospective consumers of such goods. Example: Like the mark 'M' which stands for the food items originating from the American fast food chain, McDonald. This creates an image and reputation for the food items offered by it for sale.

One may sell, bequeath or transfer the rights of a trade mark through a process called assignment. One may also licence rights to his trade mark.

6. Other Kinds of Trade Marks

The following are the other types of marks:

1. Certification mark   2. Service mark   3. Collective mark

6.1 Certification Mark

It is a mark capable of distinguishing the goods or services in connection with which it is used in the course of trade that are certified by the proprietor of the mark in respect of origin, material, mode of manufacture of goods or performance of services, quality, accuracy or other characteristics from goods or services not so certified and registrable. It can be any word, name, symbol, device, or any combination, used, or intended to be used, in commerce with the owner's permission by someone other than its owner. These marks are not to be used independently but used in conjunction with trade marks on the goods. The agency certifying the goods does not carry out the trade in goods or service. For example AGMARK, ISI, FPO, WOOLMARK, ISO, etc. are certification marks. Internationally, the certification marks become important with standardization of quality e.g. ISO. The objective of establishing these marks was to develop quality management systems that would harmonize the exchange of goods and services in the global market.

6.2 Service Mark

Service means service of any description which is made available to potential users in connection with business of any industrial or commercial matters such as banking, communication, education, financing, insurance, chit funds, real estate, transport, storage, material treatment, processing, supply of electrical or other energy, boarding, lodging, entertainment, amusement, construction, repair, conveying of news or information and advertising. A service mark is any word, name, symbol, device, or any combination, used, or intended to be used, in commerce, to identify and distinguish the services of one provider from services provided by others, and to indicate the source of the services. For example Life Insurance Corporation (LIC), State Bank of India (SBI), etc.

6.3 Collective Mark

It is a trade mark or service mark used, or intended to be used, in commerce, by the members of a cooperative, an association (not being a partnership within the meaning of Indian Partnership Act, 1932), or other collective group or organization, including a mark that indicates membership in a union, an association, or other organization. It distinguishes the goods or services of members of an association of persons which is the proprietor of the work from those of others. For example Delhi High Court Bar Association, New Delhi Mercantile Traders Association, Automobile Association (AA) of India,

INTUC, CITU, etc. The following points should be noted for registering collective marks:

i. The collective mark is owned by an association of persons not being a partnership.

ii. The collective marks belong to a group and its use thereof is reserved for members of the group.

iii. The association may not use itself the collective mark but it ensures compliance of certain quality standards by its members who may use the collective mark.

iv. The primary function of a collective mark is to indicate a trade connection with the association or organization who is the proprietor of the mark.

7.   Well-known and Associated Trademarks

Well-known trade mark in relation to any goods or services means a mark which has become so common to the substantial segment of the public which uses such goods or receives such services and the use of such mark for other goods or services would indicate a connection to the first-mentioned goods or services. Associated trade marks means trademarks deemed to be, or required to be, registered as associated trade marks under the Act. Mercedes-Benz is a well known mark in automobiles. A company started using Benz as a trademark for innerwear. Mercedes-Benz won the case against that company on the basis of well known mark though that was used in different class.

Case Studies on Well known Marks

Tata: In a case filed in 2011 by Tata Sons against Durga Scale Company, at the Delhi High Court, Tata Sons had sought to restrain Durga from engaging in the business of manufacturing and /or selling and/or directly or indirectly dealing in goods bearing the trademark A-ONE TATA. The defendant is a company engaged in manufacturing weighting scales and springing balances under the trademark A-ONE TATA. The Delhi High Court after hearing the matter of merits decided in favour of the plaintiffs (Tata Sons) establishing that TATA is a well known and household name in the Indian market and households. The question of disputing its well known status does not arise. In wake of its favourable judgement the court also awarded the amount of two lakh INR in favour of the plaintiff towards punitive damages and has restrained the defendants from mark A-ONE-TATA in respect of any goods and/or services in commerce.

Tata Sons won a case against Balbir Vohra and others over infringement of the trademark 'Tata'. The Delhi High Court held that the brand' Tata' is known worldwide and is associated with the products of Tata, while the defendants had been using the brand name 'Tata' for buckets and washers "only to encash on the goodwill and reputation of Tata Sons." The Court granted a permanent injunction against the 'Tata' trademark.

SONY: Japanese electronic giant Sony Corporation has won a trademark dispute with a Kolkata based manufacturer of lingerie and other innerwear that sold its products under the name 'abt SONY'. The Deputy Registrar of Trademarks, Kolkata, had in 2007 allowed the manufacturer to use trademark. Sony later filed an appeal before the Intellectual Property Appellate Board, with the company's counsel submitting that the Sony brand name was "internationally known and registered in more than 200 countries" as a well known mark. The counsel said his client set up an Indian subsidiary Sony India Pvt Ltd and through ancillary companies Sony Music Entertainment Incorporated and Sony Picture Entertainment, had established a worldwide music and image based brand. AB Textiles said it had been using the name 'abt Sony' for hosiery and other innerwear products in West Bengal and Assam since 1998. It said it had never used the trademark for electronic goods. The appellate board censured the decision of deputy registrar and restrained the use of Sony trademark by innerwear manufacturer.

8. Registration of Trademark

Only the owner of a trade mark may file an application for its registration. Generally, the person who uses or controls the use of the mark, and controls the nature and quality of the goods to which it is affixed, or the services for which it is used, is the owner of the mark such as companies, individuals, partnerships, trade unions or lawful associations. Trade mark registration usually involves:

1. Any person who claims to be the proprietor of a trade mark and is desirous of registration of the mark shall apply to the Registrar within whose territorial limit is the principal place of business in India of the applicant.

2. Full name, address & nationality of applicant.

3. An appropriate application form and application fees.

4. Graphic representation of the trademark. In the case of 3D mark, the reproduction of the mark should consist of a two dimensional or photographic reproduction. Drawings of the trade mark, if application is made for a word or words in special form or a design. Where the trade mark contains a word or words in scripts other than Hindi or English, a transliteration and translation of each word in English or in Hindi should be given.

5. A preliminary search (done by you or your agent) of existing trade marks.

6. Examination of application by the Trade-Marks Office after which the application is either accepted or accepted subject to amendments.

7. Publication of the application in the Trade Marks Journal.

8. Any person may, within three months from the date of the advertisement, give notice in writing to the Registrar of opposition to the registration.

9. The Registrar shall serve a copy of the notice of opposition to the applicant for registration.

10. Within two months from receipt of the notice of opposition the applicant shall send a counter statement, otherwise the application would be deemed to be abandoned.

11. A copy of notice is also sent to the person giving notice of opposition.

12. Registrar then calls both parties with their evidence.

13. If the decision goes in favour of the applicant then the trade mark is registered.

14. If decision goes against the applicant then he is free to appeal in the high court.

In general, the following marks may not be registered: i) words those are clearly descriptive (e.g. "delicious" ice cream); ii) terms those are misleading; iii) words that designate a place of origin (e.g. 'Atlantic' cold, 'Nagpur' oranges); iv) terms or symbols that are too similar to an existing trade mark (e.g. Colgate – Collegiate, Safal-Sakal, Lakme-Likeme); v) terms and symbols that are expressive in nature are prohibited under the Trade Marks Act. It includes symbols of national and international organizations and terms that are considered immoral or offensive; vi) Other types of marks that may not be registered are plant variety denominations and protected geographical indications.

When an application is made for the registration of a trade mark which falsely suggests a connection with any living person, or a person whose death took place within twenty years prior to the date of application of registration of the trade mark, then a written consent of the living person or the legal representative of the deceased person is required to be furnished.

Once you have filed an application for registration of trademark, the "TM" symbol may be used with the mark. Anyone who claims rights in a mark may use the TM (trademark) designation with the mark to alert the public to the claim. However, the registration symbol, ®, may only be used once the mark is actually registered in the Trademark Registrar's Office. Even though an application is pending, the registration symbol may not be used before the mark has actually become registered.

Globalization of trade and industry the "trans-border" reputation of a foreign trade mark has recently been recognized by various courts in India. The Indian Trade Marks Act makes no distinction between foreign and Indian brand names and advertisement of foreign mark in India is sufficient to establish the user in India.

9. Duration

The duration of a registered trade mark is for a period of ten years. The 10 year period of registration is taken from the date of making of the application which is deemed to be the date of registration. It can be renewed for next ten years from the date of expiration of the original registration by making an application in a prescribed manner within a prescribed period and subject to payment of the prescribed fees. For filing new application for trademark there is a fee of Rs 2500/-. For renewal of registered trademark fee of Rs 5000/- is applicable. Thus the trademark can be renewed perpetually. In case, the proprietor fails to renew his trade mark or is not used by the owner, Registrar may remove the trade mark from the register within one year from the expiration of last registration. It can be restored only after one year of removal from the register with condition that there has been no bona fide user of the trade mark, which has been removed during the two years immediately preceding its removal.

Rights of registration of trade mark are valid in a country where it was registered. If the products are sold in other countries, one should consider applying for foreign registration. However, there is a provision of international registrations also. Since, India is a member of the WTO and Paris Convention; it has reciprocal arrangements with their member countries. Application should be made within six months of the first application to the desired member country with certified copy of application.

Guidelines have been issued by India's Patents and Trademarks office with respect to the working of the Madrid protocol. This system will enable the firms to pay a single registration fee to be paid in India and will save them on the expenditure incurred in filing separate international trademark applications for all the individual countries.

10. Opposition to Registration

Any person can give notice in writing for opposition to the registration within three months from the date of advertisement or re-advertisement of an application for registration. Registrar will serve a copy of notice to the applicant and within two months from the receipt by the applicant of such copy of notice of opposition. The applicant will send a counter statement of the grounds on which he relies for his application. If he fails to do so the application will be abandoned there off.

11. Economic Benefits

Trade marks serve the following economic functions:

1. They facilitate consumers' decision-making about their choice of products in the market.

2. They provide incentives for an enterprise to invest in development and delivery of goods and services with the qualities as per consumers' desire.

   These two functions are complementary and mutually reinforcing. When customers chose a product because of the qualities suggested by its mark, and when businesses invest in quality to continue to build brand reputation, the result should be improved quality that yields customer loyalty to the brand.

3. An established trade mark can: i) Increase unit sales; ii) Cement customer loyalty; iii) Assist in response to competitive pressure iv) Increase revenues and profitability; v) Expand and maintain market share; vi) Differentiate products; vii) Help introduce new product lines; viii) Gain royalties through licensing programmes; ix) Provide the foundation for franchises; x) Support strategic partnerships and marketing alliances; xi) Justify corporate valuation in financial transactions xii) Raise awareness of charitable causes; xiii) Signal compliance with safety requirements; xiv) Show fulfillment of technical specifications.

4. Constructive notice nationwide of the trade marks owner's claim.

5. Evidence of ownership of the trade marks.

6. Jurisdiction of courts may be invoked.

7. Registration can be used as a basis for obtaining registration in foreign countries.

8. Registration may prevent importation of infringing foreign goods.

12. Classes of Trademarks

The India Trademarks Act, 1999 follows an international system of Nice Classification under Schedule IV enlisting the classes under which various goods and service may be applied for trademark registration. This Schedule IV under the Act has been amended to bring the same in conformity with the Ninth Edition of the Nice Classification. The numbers of classes under the Schedule have been increased from 42 to 45. There are 34 classes of goods and 11 classes of services. Earlier there were 34 classes of goods and 8 classes of services. Class 42 earlier covering 'Other services' has been replaced to include the following: 42 – Scientific and technological services and research and design relating thereto; Industrial analysis and research services; design and development of computer hardware and software; 43 – Services for providing food and drink; temporary accommodation; 44 – Medical services, veterinary services, hygienic and beauty care for human beings or animals; agriculture, horticulture and forestry services; 45 – Legal services; security services for the protection of property and individuals; personal and social services rendered by others to meet the needs of individuals. A person can make a single application for registration of a trade mark for different classes of goods and services by paying prescribed fees for each such class of goods and services.

13. Registered v/s Unregistered Trademark

A registered trade mark has been approved and entered in the Trade Marks Register held by the Trade Marks Office. Registration is proof of ownership in a court of law. An unregistered trade mark may also be recognized through common law as the property of the owner, depending on the circumstances. A registered as well as unregistered trade mark is assignable and transmissible with or without the goodwill of the business concerned. Assignments or transmission of a trade mark is restricted where multiple exclusive rights would be created in more than one person, which would be likely to deceive or cause confusion.

14. Infringement of Trademark and Penalties

Infringement of trademarks takes place when the offences are related to falsification of trademark or falsely apply the trademark to goods or services. The trademark is infringed when whole of the mark or essential feature of the mark is taken and then a few additions and alterations are made which make the mark identical or deceptively similar to the registered trade mark.

There are provisions for penalty if any one:

 i. Falsify any trade mark.

 ii. Falsely applies a trade mark to goods or services.

 iii. Makes, disposes of, or has in possession, any die, block, machine, plate or other instrument for the purpose of falsifying a trade mark.

 iv. Applies any false trade description to goods or services.

 v. Gives a false indication of country, place, name and address of the manufacturer or person for whom the goods are manufactured.

 vi. Tampers, alters or effaces an indication of origin which has been applied to any good.

The punishment for the above offences as per the law shall not be less than 6 months imprisonment which may extend to three years and a fine which shall not be less 50,000/-, but may extend to rupees two Lakhs. Wherever the court proposes a lower punishment than the minimum, it has to record, adequate and special reasons for the same. However second and subsequent offences shall be more severely punished. There shall be an imprisonment of not less than one year which may extend to three years and a fine which shall not be less rupees one lakh, but may extend to rupees two Lakh. The court can propose a lower punishment than the minimum, only after recording adequate and special reasons for the same.

Few case studies are given in Box 2.

Box 2

Lakme v/s Like-me: Claimant was selling cosmetic products under the registered trade mark "Lakme". Defendant was using the trade mark "like-me" for the same class of product. These two words have a striking resemblance and are also phonetically similar. There is a possibility of deception and confusion being created in the minds of prospective buyers of claimant's products. Injunction was made permanent.

Colgate v/s Collegiate: Claimant was a reputed manufacturer of dental cream Colgate. Defendants used a trade mark collegiate. These are phonetically similar. Defendant's mark has deceptively similar letters in white with red background as to that of claimant's mark. This would have created confusion in the minds of customer and to pass off its products as Colgate. Hence the mark was restrained through injunction.

Honda v/s Hongda: Honda Motor Co. won a record compensation award in an intellectual property infringement case when a Beijing court ordered a major local motor cycle manufacturer to pay about 1.47 million Yuan in damages for placing 'Hongda' logos on its products in May 2005.

The infringing mark is used in a regular trade wherein the proprietor of the mark is engaged. Few case studies are given in Box 3.

Box 3

Cable News Network LP (CNN) v. Cam News Network Limited: The plaintiff Cable News Network LP (CNN) is a leading news group in disseminating world news on electronic media and owns and runs the 'CNN International'. The defendant 'Cam News Network Today' is also engaged in the business of printing and publication of news, journals and magazines. Both parties herein have locked horns on the use of mark 'CNN'. The court thus restrained the defendant Cam News Network Today from publishing, selling, marketing, advertising any magazine or providing news services in any format whatsoever under the plaintiff's trade mark CNN.

Castrol case: Delhi High Court in a Court case in the matter between Castrol Limited & Other v/s Mr. Rajinder Kumar Gupta & Others found involvement and duplicate packaging of counterfeit Products viz. oils, lubricants, greases, etc., being sold in the market bearing the plaintiffs (Castrol Limited) identical trademark 'CASTROL' in 2011. To the strong allegations lodged by the Plaintiffs, the defendants however neither retorted nor negated the allegations nor did they contest the matter and were a no show even in court. The Delhi High Court relied upon the evidence placed before it by the Plaintiffs and not only did the plaintiffs receive a permanent injunction in their favour, the Delhi High Court supplemented the injunction order with a direction to the defendants to pay to the tune of 10 lakh rupees. As part of the judgment the defendants have been directed to pay interest on the aforesaid amount at the rate of 9 percent per annum from the date of filing of the suit till realization.

Original Choice v/s Officer's Choice: In a long pending trade mark dispute between Allied Blenders and Distilleries Ltd., (ABDL) proprietor of the mark 'Officer's Choice' and John Distillers Ltd., (JDL) of the mark 'Original Choice', the Intellectual property Appellate Board IPAB) has held that the mark 'Original Choice' of JDL is not deceptively similar to the mark 'Officer's Choice' of ABDL. ABDL has been opposing the mark Original Choice of JDL for several years on the ground inter alia that it was similar to its mark Officer's Choice and further since the acronym OC of both the marks was likely to create confusion. Taking into account the disclaimer over the word choice in respect of AVDL's trade mark registration for Officer's Choice, the IPAB in 2013 held that ABDL does not have exclusive right to the word choice. Furthermore, it noted that the marks involved were only words and there were no identical labels associated in the dispute. Hence the IPAB did not find any scope for confusion between the marks.

Everest, Bisleri fight over term 'Himalayas': MT EVEREST Mineral water, a Tata company, is fighting a legal battle with Bisleri International over the use of the word Himalaya in its mineral water branding. The company has approached the Delhi High Court seeking the word Himalaya or Himalayan as a trademark. Mt Everest, the Tata group company, has asked the court for an interim order restraining the defendants (Bisleri International) from using these words in relation to bottled water. While no final order has been passed the two parties have agreed to an interim arrangement till disposal of the application. As part of the interim arrangement, Bisleri has agreed to stop using the domain www. bislerihimalayan.com. It will continue marketing bottles that say "from the Himalayas" as a descriptive word in small font to denote the origin of the water. It has agreed not to use the word Himalayas in larger font. Bisleri has agreed to inform their stockists about these arrangements.

Trademark is infringed when it is printed or usually represented in advertisements, invoices, or bills. The trademark is infringed when the use is in such a manner as to render it likely to be taken as being used as trade mark. But, any oral use of the trade mark is not infringement.

The claimant in case of an infringement may be:

i. The proprietor of registered trade marks or his legal successor.

ii. A registered user of a trade mark subject to prior notice to his registered proprietor and consequent failure of the registered proprietor to take any action against the infringer.

iii. An applicant for the registration of trade marks.

iv. Legal heirs of the deceased proprietor of a trade mark.

v. Any one of the joint proprietors of a trade mark.

vi. A foreigner who is the proprietor of a trade mark registered in India, if the infringement has occurred in India.

There are provisions of penalty for applying a false trade mark, trade description, etc. The office will refuse to register a mark if it does not function as a trade mark. Not all words, names, symbols or devices function as trade marks. For example, matter, which is merely the generic name of the goods on which it is used, cannot be registered. Additionally, there are several grounds for refusal which may be summarized as:

i. The proposed mark consists of or comprises immoral, deceptive, or scandalous matter.

ii. The proposed mark may disparage or falsely suggest a connection with persons (living or dead), institutions, beliefs, or national symbols, or bring them into contempt or disrepute.

iii. The proposed mark consists of or comprises the flag or coat of arms, or other insignia of a country, or of any state or municipality, or of any foreign nation.

iv. The proposed mark consists of or comprises a name, portrait or signature identifying a particular living individual, except by that individual's written consent; or the name, signature, or portrait of a deceased person during the life of his widow, if any, except by the written consent of the widow.

v. The proposed mark so resembles a mark already registered and use of that mark on applicant's goods or services are likely to cause confusion, mistake, or deception.

vi. The proposed mark is merely descriptive or deceptively mis-descriptive of applicant's goods or services.

vi. The proposed mark is primarily geographically descriptive or deceptively geographically mis-descriptive of applicant's goods or services.

vii. The proposed mark is primarily merely a surname.

14.1 Case Studies on Trade Mark Infringement

'rehydrate, replenish, refuel' – cannot be monopolized: Heinz India wins trademark case against Pepsi Co: Heinz India, maker of Complan and Glucon-D, has won a trademark case against PepsiCo over use of a tagline for its energy drink Glucon-D Isotonik. The two firms were entangled in a bitter legal battle in the Delhi High Court over use of similar taglines for their respective energy drinks. Pepsi's Gatorade sports drink uses the tagline 'rehydrate, replenish, refuel', while Heinz advertisements say 'Glucon-D Isotonik rehydrates fluids, replenishes vital salts and recharges glucose'. Anuradha Salhotra, of Lall Lahiri & Salhotra, an IP law firm that represented Heinz in the matter pointed out that the words rehydrate, replenish and refuel are commonly used to describe the character and quality of an isotonic drink and no one can therefore claim monopoly over them. Gatorade and Glucon-D Isotonik of PepsiCo is a premium 'sports' drink targeted at adult consumers while Complan and Glucon-D the powder-based Glucon-D of Heinz targets families and children.

Low Absorb", "LOSORB" and "Lo-SORB" are descriptive and cannot be monopolised: In a matter before the Hon'ble Supreme Court in 2011, the case of *Marico Industries*, the maker of cooking oils Saffola and Sweekar was up for final hearing where Marico owner of trade mark slogans such as "Lo-Sorb", "Low Absorb" and Losorb had challenged the use of the slogan "Low absorb" by Agro Tech Foods, owner of brand of edible oil, Sundrop. Marico claims that it is the registered proprietor of the abovementioned trade marks since 2001, filed the present lawsuit in the Delhi High Court alleging it has been using the subject trade mark for its popular brands Saffola and Sweekar since beginning of 2001 for products which had the quality of absorbing less oil. Reiterated, the technology involved was based on an additive used in the production of edible oil which retards the foaming of oil during the process of frying and cooking. Agro Tech. took strong opposition to Marico's claims and was victorious at the Delhi High Court which rejected the claims of exclusivity and held that *prima facie*, no one can monopolise the mark, "Low Absorb". The Court further said that marks "Low Absorb", "LOSORB" and "Lo-SORB" are *prima facie* descriptive with no demonstrable acquired secondary distinctiveness. The matter was then appealed to the Supreme Court, where the Hon'ble Court has rejected the appeal and upheld the decision of the Delhi High Court in the matter.

Udhayam' Popular Ghee Brand Loses Trademark: The Intellectual Property Appellate Board cancelled the trademarks 'Udhayam' and 'GRB's Udhayam' of a popular ghee brand in the southern States in an order passed by the board in 2014. The board passed the order after hearing a rectification application filed by Shri Lakshmi Agro Foods Pvt Ltd, a Chennai-based company that sells dhal and other groceries under the brand name 'Udhaiyam'. While GRB Dairy Food Pvt Ltd had in recent years used the Udhayam brand name to market ghee. The counsel for GRB Dairy Food contended that they had registered 'GRB's Udhayam Ghee' as a trademark in 1993. The order passed by IPAB rejected the arguments of GRB Dairy Food and directed the Registrar, Trademarks Registry, to cancel both the trademarks 'Udhayam' and 'GRB's Udhayam'.

"Natural" – Case of Naturals ice cream: The Bombay High Court in 2011 has passed an interim order in favour of Mr. R.S. Kamath, the proprietor of Naturals ice cream against Tanco Enterprises and Kshitij Ice Creams restraining them from using the trade mark "naturals". The matter was brought before the Bombay High Court in the matter of infringement and passing off in respect of "natural ice creams" under class 30 and the domain name www.naturalicecreams.in for that intellectual property is in the name of Mr. R.S. Kamath, owner of the "Naturals" ice cream parlours, a business set up by Mr. Kamath in the year 1984, The interim order brings in its wake legal protection and relief to the brand name "naturals" as well as establishing its brand value in most certain term i.e. popularity and reputation in India.

Liv.52 – Himalaya Drug wins trademark battle: Pharma major Himalaya Drug Co. won a 15-year-old legal battle in 2012 over its medicine Liv.52, with the Delhi High Court holding that homoeopathic firm SBL Limited infringed the trademark by coming up with its own preparation named as Liv-T. "The defendant (SBL Ltd) is restrained from using the mark LIV as part of its trade mark LIV-T while dealing with the medicinal preparation. The court set aside its single judge bench judgement which had held that homoeopathic firm SBL Limited did not infringe trademark Liv.52 of Himalaya Drug Company, registered way back in 1957, in making its own medicine Liv-T. Writing the verdict for the bench, Justice Manmohan Singh granted six months to SBL to "liquidate" existing stock of Liv-T."It arrived at the finding that the 'LIV' written in isolation is an essential feature of the trademark Liv.52 and also noticed the rule of comparison which is that the marks are to be compared as a whole. Therefore, presence of mark 'LIV' which is an essential feature of the mark Liv.52 shall be considered for the purposes of comparison with that of LIV-T." the court said.

Maggi to withdraw Cuppa Mania noodles: One of the Nestle India's biggest brands—Maggi- has landed in a soup-literally. Maggi's variant- Cuppa Mania noodles in two variants Masala Yo and Chilli Chow Yo were withdrawn barely in two months after a high profile launch in 2008. In reaction to complaint filed by Moods Hospitality, the Delhi High Court has ordered Nestle to restrain from all sales, manufacturing, marketing and advertising activity for its Masala Yo and Chilli Chow Yo variants till further notice. The court has ordered Nestle India to withdraw all activity for the product across the country. Moods Hospitality had filed the case against Nestle in the Delhi High Court in May 2008. The case alleges that Nestle's noodles in a cup are a direct take off on 'Yo! China's and 'Yo! on the Go' noodles. Stating that Yo! is synonymous with Chinese noodles, Moods Hospitality had alleged that Nestle's act amounts to trademark infringement, and that Nestle's new launches are a direct take off on its Yo! on the Go product.

Trademark protection to Colour themes for T-shirts: Louisiana State, Ohio State, Southern California and the University of Oklahoma sued Smack Apparel Company alleging that the T-shirts sold by the company with the Universities' colour schemes violated the trademarks of the Universities. The Federal Court held that copying of the colour schemes of the T- shirts by the company amounts to passing off under the trademark law because such T-shirts are aimed at gaining profits by establishing a link with the Universities' products. The court pointed out that the company would be liable even if it had not used the names, logos or captions of the Universities. The Universities were granted damages of about forty six thousand US dollars.

15. Passing Off

Passing off action arises when an unregistered trade mark is used by a person who is not the proprietor of the said trade mark in relation to goods or services of the trade mark owner. Passing off in India is a tort actionable under the common law and mainly used to protect the goodwill attached with unregistered trade marks. Plaintiff is required to establish deceptive similarity of contesting marks and also to prove deception or confusion among public and likelihood of injury to plaintiff's goodwill. Regular rules of jurisdiction provided under section 20 of Civil Procedure Code 1908 applies. Passing off action can only be instituted at the level of district court where defendant resides or carries on the business or cause of action has arisen.

Whereas trademark is said to be infringed when a registered trademark is used by a person who is neither the registered proprietor nor the licensee of trademark in relation to goods or services for which it is registered.

16. Domain Name

It is a name owned by a person or organization and consisting of alphabetical or alphanumeric sequence followed by a suffix indicating the top level domain: used as an internet address to identify the location of particular web pages. The named address includes:

- The type of organization
- The name of the organization
- Country of origin

Thus a domain name is an identification string that defines a realm of administrative autonomy, authority or control within the Internet. Domain names are formed by the rules and procedures of the Domain Name System (DNS). Any name registered in the DNS is a domain name. Domain names can also be thought of as a location where certain information or activities can be found. In general, a domain name represents an Internet Protocol (IP) resource, such as a personal computer used to access the Internet, a server computer hosting a web site, or the web site itself or any other service communicated via the Internet. In 2014, the number of active domains reached 271 million.

A domain name consists of one or more parts, technically called *labels*, that are conventionally concatenated, and delimited by dots, such as *example.com*, *yourdictionary.com*. Labels in the DNS are case-insensitive, and may therefore be written in any desired capitalization method, but most commonly domain names are written in lowercase in technical contexts. The right-most label conveys the top-level domain; for example, the domain name *www.example. com* belongs to the top-level domain *com*. The top-level domains (TLDs) such as com, net and org are the highest level of domain names of the Internet. An example of a domain name showing a country of origin is *.ca* for Canada or *.uk* for the United Kingdom or *.in* for India.

The registration of these domain names is usually administered by domain name registrars who sell their services to the public. Today, the Internet Corporation for Assigned Names and Numbers (ICANN) manages the top-level development and architecture of the Internet domain name space. It authorizes domain name registrars, through which domain names may be registered and reassigned.

Domain names are used to establish a unique identity. Organizations can choose a domain name that corresponds to their name, helping internet users to reach them easily. A generic domain is a name that defines a general category, rather than a specific or personal instance, for example, the name of an industry, rather than a company name. Some examples of generic names are *books.com*, *music.com*, and *travel.info*. Companies have created brands based on generic names, and such generic domain names may be valuable.

Domain names are often simply referred to as *domains* and domain name registrants are frequently referred to as *domain owners*, although domain name registration with a registrar does not confer any legal ownership of the domain name, only an exclusive right of use for a particular duration of time. The use of domain names in commerce may subject them to trademark law.

16.1 Case Studies on Domain Name

WIPO evicts cyber squatter form India's SBI domain: An Australian entity, which hijacked the domain name of State Bank of India (SBI) cards hoping to later sell it for a hefty sum to the State Bank of India subsidiary, has been evicted by WIPO. The WIPO Administrative Panel, based in Geneva, in a decision in 2005 held that the Australian entity had constructed a website in the domain name sbicards.com 'in bad faith' and ordered transfer of the domain name to the SBI, Indian company.

Google wins Domain Name case: Internet search giant, Google has won a cyber squatting case at the WIPO in 2009 against an Indian who tried to block the domain name 'googblog.com'. According to the information available with the WIPO, Geneva-based WIPO Arbitration and Mediation Center has ordered the transfer of domain name to the US-based search giant after Herit Shah of Gujarat offered to surrender the disputed name to Google. Google had challenged the registering of domain name 'googblog.com' by Shah at WIPO stating that it was confusingly similar to its trademark on which the company has rights. Cyber squatting is an illegal activity of buying and officially recording an address on the internet that is the name of an existing company or a well-known person, with the intention of selling it to the owner in order to make money.

Mahindra & Mahindra Wins Cybersquatting Case at WIPO: Mahindra & Mahindra has won a cybersquatting case at WIPO against a US based firm in 2009, which had tried to block domain name, 'mahindralogistics.com'. M&M had challenged registering of the domain name by Portfolio Brains LLC of the US, at the WIPO stating that it was identical and confusing with its subsidiary, Mahindra Logistics Ltd. and "deceptively similar to the Mahindra marks". Geneva-based WIPO Arbitration and Mediation Centre has ordered the transfer of domain name to M&M. The disputed domain name "mahindralogistics.com" is registered with Nameking.com by Portfolio Brains. M&M had filed the complaint against the US based firm on 18th Feb., 2009, although the domain name was registered on 12th March, 2007.

WIPO sides in Favour of "Jenny from the Block' in Domain Matter: WIPO in 2009 ordered that domain names jenniferlopez.net and jenniferlopez.org registered in the name of a Jeremiah Tieman be transferred to the Jennifer Lopez website after it was found that the websites were not truly fan web sites, but were linked to a web site that generated revenues through paid advertisements.

WIPO gives Domain name indianoil.org to Indian Oil: In a domain name related legal battle over the well-known trade mark Indian Oil, the World Intellectual Property has sided in favour of the Indian Oil Corporation, India. The brief facts, Mr. Nitin Jindal also owner of search engine GODaddy.com claimed ownership over the domain Indianoil.org. WIPO taking congnizance of complaint and matter lodged by Indian Oil Corporation, India recognized the well-known status of the Indian company and has accordingly granted exclusive rights of the trade mark and domain name to Indian Oil. Mr. Jindal has been directed to transfer the ownership and all related rights to the Indian company and rightful owner.

Cybersquatting case relating to the domain name www.arunjaitley.com: Arun Jaitley well known politician and leader of the BJP, wished to register the domain www.arunjaitley.com but found that the domain had already been registered as of the year 2009. The plaintiff sent a letter to Network Solutions (the 2nd defendant) requesting registration on 16th July, 2009. At the time of sending this letter to the defendant, the domain name was 'pending deletion'. The next day he received a reply asking him to purchase the domain name from them, or wait for the domain to be deleted. However, domain www.arunjaitley.com according to the WHOIS report had expired on 12th June, 2009. According to the Domain Deletion Policy, the domain ought to have been deleted after the expiration of 35 days. The defendants asked the plaintiff to purchase the domain from them at a cost of anything between $11,725 to $14,475. Thus, the plaintiff argued that monetary gain was the objective of the defendants. Further, Network Solutions thereafter proceeded to transfer the domain to Portfolio Brains, an auction site for domain names. The court rightly held that domain names are protected under the law of passing off and personal names constituting domain names would be granted similar protection. The court found the conduct of the defendants to be violative of the ICANN policy on the grounds of bad faith registration as well as insufficient cause to keep the domain name in its possession after the period of expiry. Thus, the court found malafide intention on the part of the defendants in trafficking the domain name, with the objective of extracting a hefty auction price from the plaintiff in 2011. It thus granted punitive damages to the tune of Rs.5 lakhs to the plaintiff for *"causing hardship and harassment and mental torture to the plaintiff in getting back the domain name"*. The court further restrained Portfolio Brains from adopting, or using the mark in any manner whatsoever in cyberspace. It also directed the transfer of the said domain to the plaintiff with immediate effect. It made a direction under the ICANN rules to block the domain name and immediately transfer this domain name to the plaintiff with requisite charges and formalities.

Selected reading

Chawla, H.S. and Singh, A.K. 2005. Intellectual Property Rights. Vol. II: Copyrights, Trade Marks, Trade Secrets and Geographical Indications. Pantnagar University Press, pp. 75.

http//www.ipindia.nic.in

http://www.wipo.int/treaties/en/

The Trade Marks Act, 1999 along with The Trade Marks Rules 2002, Universal Law Publishing Co., Delhi, 2011.

Industrial Design

A Design refers to the features of shape, configuration, pattern, ornamentation or composition of lines or colours applied to any article, whether in two or three dimensional (or both) forms. This may be applied by any industrial process or means (manual, mechanical or chemical) separately or by a combined process, which in the finished article appeals to and judged solely by the eye. Design does not include any mode or principle of construction or anything which is mere mechanical device. It also does not include any trade mark or any artistic work. An industrial design registration protects the ornamental or aesthetic aspect of an article. The object of the Designs Act is to protect new or original designs so created to be applied or applicable to particular article to be manufactured by industrial process or means. Sometimes purchase of articles for use is influenced not only by their practical efficiency but also by their appearance.

The important purpose of design Registration is to see that the artisan, creator, originator of a design having aesthetic look is not deprived of his bonafide reward by others applying it to their goods.

Designs are applied to a wide variety of products of different industries like handicrafts, medical instruments, watches, jewelry, house wares, electrical appliances, vehicles and architectural structures. An industrial design is primarily for aesthetic features. These are intended to encourage the decorative arts to enhance the commercial values of an article and to enlarge the commercial demand for that article.

1. International Treaties and Conventions

1.1 Hague Agreement: The International Registration of Industrial Designs

On November 6, 1925, the Hague agreement concerning the International Deposit of Industrial Designs was adopted within the framework of the Paris Convention. The Agreement entered into force on June 1, 1928, and has been revised and supplemented several times, the London Act of 1934 and The Hague Act of 1960 (referred respectively as "the 1934 Act" and "the 1960 Act") and the The Geneva Act of 1999 (the "1999 Act"). Two Acts of the Hague Agreement are currently in operation – the 1999 Act and the 1960 Act. In September 2009, it was decided to freeze the application of the 1934 Act of the Hague Agreement, thus simplifying and streamlining overall administration of the international design registration system. The 1999 Act of the Agreement is open to any WIPO Member State and to certain intergovernmental organizations. While the 1960 Act remains open to States party to the Paris Convention for the Protection of Industrial Property (1883).

The application of the 1934 Act was frozen as of January 1, 2010, meaning that no new registration or designation under the 1934 Act could be entered in the International Register as of that date. However, the renewal of existing designations under the 1934 Act and the recording in the International Register of any change affecting such designations will continue to be possible up to the maximum duration of protection under the 1934 Act (15 years).

1.2   The Locarno Agreement: International Classification for Industrial Designs

The Locarno agreement established an International Classification for Industrial Designs, is a multilateral international treaty, which was signed on October 8, 1968. It entered into force on April 27, 1971. The Locarno Classification comprises three parts:

- A list of Classes and Subclasses; in total, there are 31 classes and 211 subclasses;

- An alphabetical list of Goods in which industrial designs are incorporated; this List contains in total approximately 6,000 entries;

- Explanatory notes.

According to Article 2(3) of the Locarno Agreement, the industrial property offices of the countries of the Locarno Union must include in the official documents for the deposit or registration of designs, and if they are officially published, in the publications in question, the numbers of the classes and subclasses of the Locarno Classification in which the goods incorporating the designs belong. Each country may attribute to such classification the legal consequences, if any, which it considers appropriate. The Locarno Classification does not bind the countries of the Locarno Union as regards the nature and the scope of protection afforded to the design in those countries. Further, Locarno agreement provides that the countries of the Locarno Union are free to adopt the Locarno classification as the only classification to be used for industrial designs, or to maintain an existing national classification system for industrial designs and to use the Locarno Classification as a supplementary classification, also to be included in official documents and publications concerning the deposit or registration of designs.

2. Design Law in India

The Designs Act, 2000 and the Designs Rules, 2001 presently govern the design law in India. The Act came into force on 25th May 2000 while the Rules came into effect on 11th May 2001. The present legislation is aligned with the changed technical and commercial scenario and made to conform to international trends in design administration.

In most countries, an industrial design needs to be registered in order to be protected under industrial design law as a "registered design". In some countries, industrial designs are protected under patent law as "design patents".

Industrial design laws in some countries grant – without registration – time and scope limited protection to so-called "unregistered industrial designs". Depending on the particular national law and the kind of design, industrial designs may also be protected as works of art under copyright law.

3. Registration

Under the Designs Act, 2000 the "article" means any article of manufacture and any substance, artificial, or partly artificial and partly natural; and includes any part of an article capable of being made and sold separately (e.g. cup and saucer). First-to-file rule is applicable for registrability of design. If two or more applications relating to an identical or a similar design are filed on different dates, first application will only be considered for registration of design.

3.1 Essential Requirements for Registration of Design

- A design should be new or original.

  A design should not be disclosed to the public anywhere by publication in tangible form or by use or in any other way prior to the filing date. The novelty may reside in the application of a known shape or pattern to new subject matter. It must significantly differs from known designs or combinations of known design features.

- Considering the novelty and/or originality requirement for industrial designs in most legislations which varies for country to country, it is in general crucial to file an application for registration or for the grant of a design before publicly disclosing it, so as to avoid destroying its novelty/originality. If the industrial design has already been disclosed to the public (for example, by an advertisement published on the company's website), it may no longer be considered as "new" or "original" and may become part of the public domain. Some countries, however, allow for a "grace period" of 6 or 12 months to file after disclosure of the industrial design.

- India is one of the countries party to the Paris Convention so the provisions for the right of priority is applicable. On the basis of a regular first application filed in one of the contracting state of the party to Paris Convention (e.g. 4th Jan., 2015), the applicant may within the six months apply for protection in other contracting states (e.g. 10th May, 2015). The application will be regarded as if it had been filed on the same day as the first application (4th Jan., 2015).

- The design should relate to features of shape, configuration, pattern or ornamentation applied or applicable to an article. Thus, designs of industrial plans, layouts and installations are not registrable under this Act. They can be registered under copyright.

- A design should be significantly distinguishable from known designs or combination of known designs.

- The design should be applied or applicable to any article by any industrial process. Normally, designs of artistic nature like painting, sculptures and the like which are not produced in bulk by any industrial process are excluded from registration under the Act. Paintings and sculptures are subject matter of copyright.

- The design should not comprise or contain scandalous or obscene matter.

- A design should not to be a mere mechanical contrivance. It means that any mode or principle of construction or operation or anything which is in substance a mere mechanical device, would not be registrable under the Designs Act. For example a key having its novelty in the shape of corrugation or bend at the portion intended to engage with levers inside the lock associated with, cannot be registered as a design under the Act.

- Design as applied to an article should be integral with the article itself. Because once the design i.e. ornamentation is removed then only a piece of paper, metal or like material remains and the article ceases to exist.

- A design should be applied to an article and should appeal to the eye. This means that design must appear and should be visible on the finished article. Thus, any design in the inside arrangement of a box or almirah or purse may not be considered as design, as these articles are generally put in the market in the closed state.

- A design should not be contrary to public order or morality.

- The design should not include any trademark or property mark or artistic work.

3.2 Applicant

Any person or the legal representative or the assignee can apply separately or jointly for the registration of a design. The term "person" includes firm, partnership and a body corporate. An application may also be filed through an agent in which case a power of attorney is required to be filed. An Application for registration of design may be prepared either by the applicant or with the professional help of attorneys. The Design Wing of the Patent Office may be approached for finding out whether a design has been previously registered or not on prescribed form. Before filing an application, the applicant can also obtain information whether the design has already been registered or not, by filing a request in Form along with prescribed fee.

3.3 Procedure for Submission of Application of Registration

Any person who desires to register a design is required to submit the following documents to the Design Wing of the Patent Office.

i. Application duly filled in on the prescribed form along with the prescribed fees, stating name in full, address, nationality, name of the article, class number, address for service in India. The application should also be signed either by the applicant or by his authorized agent.

ii. Representation (in quadruplicate of size 33 cm × 20.5 cm with a suitable margin) of the article. Drawings/sketches should clearly show the features of the design from different views and state the view (e.g. front or side).

iii. A statement of novelty and disclaimer (if any) in respect of mechanical action, trademark, work, letter, numerals should be endorsed on each representation sheet which should be duly signed and dated. [The following statement of novelty should be mentioned on the representation of a design as per the Act: "The novelty resides in the shape and configuration of the article as illustrated." "The novelty resides in the portion marked as 'A' and 'B' of the article as illustrated." "The novelty resides in the ornamentation or surface pattern of the article as illustrated." For example: The novelty resides in the floral ornamentation of the carpet as illustrated. If the ornamental pattern on an article is likely to be confused with a trade mark, a disclaimer may be made in the following manner:- No claim is made by virtue of this registration to any right to the use as a trade mark of what is shown in the representations.

iv. If the representation suggests any mechanical action of the article a disclaimer may be inserted in the following manner:- No claim is made by virtue of this registration in respect of any mechanical or other action of the mechanism whatever or in respect of any mode or principle of construction of the article.

v. If the representation contains words, letters, numerals, etc., a disclaimer may be inserted in the following manner:.- No claim is made by virtue of this registration to any right to the exclusive use of the words, letters, numerals, flags, crowns, etc. appearing in the design.

   vi. Power of attorney (if necessary).

  vii. Priority documents (if any) in case of convention application claimed under Section 44 of the Designs Act, 2000.

It is not mandatory to make the article by industrial process or means before making an application for registration of design. Since, design means a conception or suggestion or idea of a shape or pattern which can be applied to an article or intended to be applied by industrial process or means. For example- a new shape which can be applied to a pen capable of producing a new visual appearance of a pen, it is not mandatory to produce the pen first and then make an application.

     The registration of a design confers upon the registered proprietor the exclusive right to apply a design to the article in the class in which the design has been registered. He can sue for infringement, if his right is infringed by any person. He can license or sell his design as legal property for a consideration or royalty.

4.   Exclusions from Scope of Design

Designs that are primarily literary or artistic in character are not protected under the Designs Act.
These will include:

- Books, jackets, calendars, certificates, forms-and other documents, dressmaking patterns, greeting cards, leaflets, maps and plan cards, postcards, stamps, medals.
- Labels, tokens, cards, cartoons.
- Any principle or mode of construction of an article.
- Mere mechanical contrivance.
- Buildings and structures.
- Parts of articles not manufactured and sold separately.
- Variations commonly used in the trade.
- Mere workshop alterations of components of an assembly.
- Mere change in size of article.
- Flags, emblems or signs of any country.
- Layout designs of integrated circuits.

5.   Register of Design

The Register of Designs is a document maintained by the Patent Office, Kolkata as a statutory requirement. It contains the design number, date of filing and reciprocity date (if any), name and address of proprietor and such other matters as would affect the validity of proprietorship of the design such as notifications of assignments and of transmissions of registered designs, etc. It is open for public inspection on payment of prescribed fee and extract from register may also be obtained on request with the prescribed fee.

6. Duration

Registration initially confers this right for ten years from the date of registration. In cases where claim to priority has been allowed, the duration is ten years from the priority date. The date of registration except in case of priority is the actual date of filing of the application. It is renewable for a further period of five years on an application with requisite fee before the expiry of the said initial period. There is provision for the restoration of a lapsed design if the application for restoration is filed within one year from the date of cessation in the prescribed manner.

7. Jurisdiction of Design

Industrial design rights are territorial. This means that these rights are limited to the country (or region) where protection is granted.

At present, no "world" or "international" industrial design right exists. In order to obtain protection in other countries, an application for the registration of an industrial design or for the grant of a patent for an industrial design must be filed in each country where protection is sought, in accordance with the law of that country. In other words, if protection is sought in countries A and B, an application should be filed with the intellectual property (IP) office of country A and another application with the IP office of country B. To avoid having to submit applications in each and every country where protection is sought, WIPO's Hague System of international application provides a practical business solution for registering up to 100 designs in a large number of territories through one single international application. [See international design registration].

In certain regions, it is also possible to obtain protection for industrial designs in the region concerned by filing an application with a regional IP office. This is the case in the African Intellectual Property Organization (OAPI), which registers industrial designs in states party to the Bangui Agreement; the African Regional Intellectual Property Organization (ARIPO), which registers industrial designs in states party to the Lusaka Agreement; the Benelux Office for Intellectual Property (BOIP), which registers industrial designs in the three "Benelux" countries; and the Office for Harmonization in the Internal Market (trademarks and designs) (OHIM), which registers industrial designs in the member states of the European Union.

8. Cancellation of Registration of Design

The registration of a design may be cancelled at any time after the registration of design on a petition for cancellation in form 8 with a fee of Rs. 1,500/-to the Controller of Designs on the following grounds:

1. that the design has been previously registered in India or
2. that it has been published in India or elsewhere prior to date of registration or
3. the design is not new or original or
4. design is not registrable or
5. it is not a design under Clause (d) of Section 2.

9. Piracy of Registered Design

Piracy of a design means the application of a design or its imitation to any article belonging to class of articles in which the design has been registered for the purpose of sale or importation of such articles without the written consent of the registered proprietor. Publishing such articles or exposing terms for sale with knowledge of the unauthorized application of the design to them also involves piracy of the design. During the existence of right over any design, other persons are prohibited from using the design except or with the permission of the proprietor, his licensee or assignee. The following activities are considered to be infringement.

- To apply the design or any fraudulent imitation of it to any article for sale [See Box 1];
- To import for sale any article to which the design or fraudulent or obvious imitation of it, has been applied;
- To publish or to expose for sale knowing that the design or any fraudulent or obvious imitation of it has been applied to it.

Thus, it would be always advantageous to the registered proprietors to mark the article so as to indicate the number of the registered design except in the case of Textile designs. Without this the registered proprietor would not be entitled to claim damages from any infringer unless the registered proprietor establishes that the registered proprietor took all proper steps to ensure the marking of the article. Also the registered proprietor show that the infringement took place after the person guilty thereof knew or had received notice of the existence of the copyright in the design.

10. Penalty

If anyone contravenes the copyright in a design he/she is liable for every offence to pay a sum not exceeding Rs. 25,000/- to the registered proprietor subject to a maximum of Rs. 50,000/- recoverable as contract debt in respect of any one design. The registered proprietor may bring a suit for the recovery of the damages for any such contravention and for injunction against repetition of the same. Total sum recoverable shall not exceed Rs. 50,000/- as contract debt as stated in Section 22(2)(a). The suit for infringement, recovery of damage etc. should be filed in any court not below the court of District Judge.

Few cases of piracy of design [See Box 1].

Crocs sues Bata, others for Design Infringement

American foot-wear company Crocs has taken on Bata and several other shoemakers and retailers in India as it tries to put its foot down on design infringement in 2014. Crocs, known for its Swiss cheese style shoes, has initiated legal action against Bata, Coqui (a Chinese footwear brand distributed by DLF brands), Suncorp Exim (for Warner Bros, clogs), Relaxo and Bioworld for infringement on its legal rights on clogs. Crocs, which has filed the cases in the District Court of Delhi, has obtained interim reliefs of the injunction. Most of the cases now have been transferred to the Delhi High Court by virtue of applicable statutory provisions. The rulings have subsequently resulted in raids across various stores in Delhi and NCR. At the heart of the contention is Crocs' holed, foam like shoe that retails for around Rs 2300. The Crocs design is the most infringed upon, said GM, Crocs India. Similar shoes by Coqui and Warner Bros, available at various offline and online retailers cost less than Rs 999/-.

HC Protects Shape of Vodka Bottle

Courts now protect the bulbous-dome shape of a Vodka bottle, which is inspired by the architecture of the Russian Orthodox Church. A top premium multinational vodka brand, Gorbatschow Wodka, moved Bombay HC after an Indian rival designed bottles similar to theirs. Court restrained, the Indian company, John Distilleries, from launching its Vodka, Salute, in bottles that resembled those of the multinational brand saying a "distinctive and capricious shape" of even a bottle could not be copied as it could confuse buyers.

11. Classification of Goods

The classification of goods is based on Locarno Agreement. In the third Schedule of Design Rules, 2001 the classification of goods has been mentioned. Under the rules only one class number is to be mentioned in one particular application. The classification of goods is based upon the function of the goods. There are classes and most of the classes are further divided into sub-classes. Under the third schedule the goods have been classified into 31 classes and another class 99 as miscellaneous which include all the products not included in the preceding 31 classes. In some specific cases, the function of article is required to be mentioned along with the name of article in the application form. Normally, the name of the article should be such that is commonly familiar in the trade or industries. The name of the article as mentioned in the application form should correspond with the representation of the article as filed. Examples: If the article relates to "chair' or sofa-cum-bed', the name should be provided accordingly and should be classified under Class 06 and sub-class – 01 where it is stated as beds and seats since it is classified in function/purpose oriented manner. If the design is applied to a toothbrush it will be classified under class 04-02. Similarly if the design is applied to a calculator, it will be classified in class 18-01.

Another application by the same proprietor for registration of same or similar design applied to any article of the same class is possible, but period of registration will be valid only up to the period of previous registration of same design.

12. International Registration

The Hague agreement is an international registration system which offers the possibility of obtaining protection for industrial designs in a number of states/contracting parties. The Hague agreement allows applicants to register an industrial design by filing a single application with the International Bureau of WIPO, enabling design owners to protect their designs with minimum formalities in multiple countries or regions. The Hague agreement also simplifies the management of an industrial design registration, since it is possible to record subsequent changes and to renew the international registration through a single procedural step.

The possibility of filing an international application under the Hague agreement is open to members who are signatories to the Hague agreement and the applicant must satisfy one, at least of the conditions of: a national/ having a domicile/ having a real and effective industrial or commercial establishment in the territory of contracting party. In addition, but only under the 1999 Act, an application may be filed on the basis of habitual residence in a contracting party.

An international application may be governed by the 1999 Act, the 1960 Act or both, depending on the Contracting Party with which the applicant has the connection described above. Protection can be obtained only in those Contracting Parties which are party to the same Act. For example, if an applicant has claimed entitlement through a Contracting Party bound exclusively by the 1999 Act, he may request protection in those Contracting Parties which are bound by the 1999 Act. The Hague system cannot be used to protect an industrial design in a country which is not party to the Hague agreement. Thus, in order to protect a design in such a country, the applicant has no choice but to file a national (or regional) application.

International design applications may be filed with the International Bureau of WIPO, either directly or through the industrial property office of the Contracting Party of origin if the law of that Contracting Party so permits or requires. In practice, however, virtually all international applications are filed directly with the International Bureau, and the majority is filed using the electronic filing interface on WIPO's website. Two dimensional designs (textile designs) would be eligible for protection

International applications may include up to 100 designs, provided they all belong to the same class of the International Classification for Industrial Designs (Locarno Classification). Applicant(s) may choose to file an application in English, French or Spanish. International applications must contain one or several reproductions of the industrial design(s) together with the designations of contracting parties where protection is sought, and must designate at least one contracting party. International registrations are published in the *International Designs Bulletin,* issued weekly online. Depending on the Contracting Parties designated, applicant(s) may request that the publication be deferred by a period not exceeding 30 months from the date of the international registration or, if priority is claimed, from the priority date.

Each Contracting Party designated by the applicant may refuse protection within 6 months, or possibly 12 months under the 1999 Act, from the date of publication of the international registration. Refusal of protection can only be based on requirements of the domestic law other than the formalities and administrative acts to be accomplished under the domestic law by the office of the Contracting Party that refuses protection. If no refusal is notified by a given designated Contracting Party within the prescribed time limit (or if such refusal has subsequently been withdrawn), the international registration has effect as a grant of protection in that Contracting Party, under the law of that Contracting Party.

The term of protection is five years, renewable for at least one five-year period under the 1960 Act, or two such periods under the 1999 Act. If the legislation of a Contracting Party provides for a longer term of protection, protection of the same duration shall, on the basis of the international registration and its renewals, be granted in that Contracting Party to designs that have been the subject of an international registration. To facilitate access to the Hague system for design creators from least developed countries (LDCs), the fees for an international application are, in their case, reduced to 10 per cent of the prescribed amounts.

It must therefore be stressed that the Hague system is merely an agreement for international procedure. Any substantive aspect of protection is entirely a matter for the domestic legislation of each designated Contracting party.

India is not yet a member of the Hague Agreement and hence, the above provisions or descriptions may not be of immediate relevance.

13. Industrial Design v/s Patent

An industrial design right protects only the appearance or aesthetic features of a product, whereas a patent protects an invention of a new product or process that offers a new technical solution to a problem. In principle, an industrial design right does not protect the technical or functional features of a product. Such features could, however, potentially be protected by a patent.

Selected Reading

http//www.ipindia.nic.in
http://www.wipo.int/treaties/en
The Designs Act, 2000 along with The Designs Rules 2001, Universal Law Publishing Co., Delhi, 2011.

Geographical Indications

Every region has its claim to fame. Christopher Columbus sailed from Europe to chart out a new route to capture the wealth of rich Indian spices. English breeders imported Arabian horses to sire Derby winners. China silk, Dhaka muslin, Venetian glass all were much sought after treasures. Each reputation was carefully built up and painstakingly maintained by the masters of that region, combining the best of Nature and Man, traditionally handed over from one generation to the next for centuries. Gradually, a specific link between the goods and place of production evolved resulting in growth of geographical indications. Legal protection for use of such terms is a comparatively recent phenomenon although France, through a parliament decree in the 15th century had regulated the use of term Roquefort for cheese. **Law for the Protection of the Place of Origin** passed by the French Parliament on 6th May, 1919 is considered as the first modern legislation. This law specified the region and community where a given product should be manufactured. Subsequently, many European countries enacted various legislations regulating use of appellations of origin of products.

Geographical Indications of Goods are defined as that aspect of industrial property which refers to the geographical indication referring to a country or to a place situated therein as being the country or place of origin of that product. Typically, such a name conveys an assurance of quality and distinctiveness which is essentially attributable to the fact of its origin in that defined geographical locality, region or country. A number of treaties administered by the World Intellectual Property Organization (WIPO) provide for the protection of GIs, most notably the Paris Convention for the Protection of Industrial Property of 1883 and the Lisbon Agreement for the Protection of Appellations of Origin and their International Registration. Under Articles 1 (2) and 10 of the Paris Convention for the Protection of Industrial Property, geographical indications are covered as an element of IPRs. That convention included all manufactured or natural products such as "wines, grain, tobacco leaf, fruit, cattle, minerals, mineral water, beer, flowers and flour" within the ambit of industrial property. It, however, had not gone in to specifics of protecting indications of source except to the extent of prohibiting false or misleading indications and allegations under the article relating to unfair competition. TRIPs Agreement under Articles 22 to 24, which was part of the WTO Agreements concluding the Uruguay Round of GATT negotiations, enlarged the scope of protection and made it a commitment of member states to provide legal means to protect geographical indications.

It may be noted that in the negotiations of the WTO, Act embodying the results of the Uruguay Round of multilateral trade negotiations authenticated by 123 nations (including India) on April 15th 1994 at Marrakesh, for the TRIPs treaty, France and the UK were successful in including an additional clause under GI to cover wines and spirits particularly to protect Champagne (produced in Champagne province of France) and Scotch Whisky (produced in Scotland) so that no other manufacturer located in any other part of the world could call its whisky as scotch or wine as Champagne.

Countries like India would endeavour to protect their traditional goods, where the name of good is synonymous with a particular region where it is produced, grown or manufactured, which signifies its quality or design or method of manufacture. Food items like – Alphonso mango, Darjeeling tea, Basmati rice, Totapuri mango, Chanderi saree, Agra petha, Nagpur oranges, Allahabad surkha, Bikaneri bhujia, Kolahapuri chappals, could also be protected under GIs.

The issue of GI protection became especially important for India after Rice Tech. Inc., USA obtained a patent for basmati rice lines and grains in the United States. Likewise Nigerian and Sri Lankan Tea growers have been passing off their tea as Darjeeling Premium Tea, which commands the highest price in the market. Until recently, protection from such misuse was granted through passing off action in courts or through certification marks. However, under the TRIPs agreement, Article 22, there is a provision that each member country must provide legislation for goods originating in a particular geographic area to prevent the misuse of goods in question originated in a geographical area other than the true place of origin of the goods. India did not have any specific law to protect geographical indications when it joined the TRIPs agreement, thus it could not enforce on any one in the world that these goods are ours. Unless a GI is protected in the country of its origin, there is no obligation under the TRIPs for other countries to extend reciprocal protection. India would, on the other hand, be required to extend protection to goods imported from other countries, which provide for such protection. In order to provide better protection to goods, the Geographical Indication of Goods (Registration & Protection) Act, 1999 has been enacted in India and has come into force with effect from 15th September 2003.

1. Treaties and Conventions

1.1 The Lisbon Agreement for the Protection of Appellations of Origin and their International Registration

Lisbon Agreement was adopted in 1958 and revised at Stockholm in 1967. It entered into force on September 25, 1966, and is administered by the International Bureau of WIPO, which keeps the International Register of Appellations of Origin and publishes a bulletin entitled Appellations of origin. The Agreement is supplemented by Regulations. The latest version of these Regulations was adopted in September 2011, with a date of entry into force of January 1, 2012.

The Lisbon Agreement is a special agreement under Article 19 of the Paris Convention for the Protection of Industrial Property. Any country party to the Convention may accede to the Agreement. Countries adhering to the Lisbon Agreement (1967) become members of the Lisbon Union Assembly. Article 2(1) of the Lisbon Agreement defines an "appellation of origin" as "the geographical denomination of a country, region, or locality, which serves to designate a product originating therein, the quality or characteristics of which are due exclusively or essentially to the geographical environment, including natural and human factors". Article 2(2) defines the "country of origin" as "the country whose name, or the country in which is situated the region or locality whose name, constitutes the appellation of origin that has given the product its reputation".

2. Law Relating to Geographical Indication of Goods in India

The Geographical Indication of Goods (Registration and Protection) Act, 1999, and The Geographical Indications of Goods (Registration and Protection) Rules, 2003, provides for the registration of geographical indications (GI) of Indian goods. Both the Act and rules were brought in to force on 15th Sept., 2003. India follows a *sui generis* system to protect its GIs. The Controller-General of Patents, Designs and Trade Marks shall be the Registrar of Geographical Indications. The Geographical Indication Registry is located at Chennai.

According to the Act "Geographical Indication", in relation to goods, means an indication which identifies goods such as agricultural, natural or manufactured or any goods of handicraft or of industry, including food stuff as originating, or manufactured in the territory of country, or a region or locality in the territory, where a given quality, reputation or other characteristic of such goods is essentially attributable to its geographical origin. In case of manufactured goods, one of the activities of either the production or of processing or preparation of the goods concerned takes place in such territory, region, or locality. "Goods" means any agricultural, natural or manufactured goods or any goods of handicraft or of industry and also includes foodstuff. "Indication" includes any name, geographical or figurative representation or any combination of them conveying or suggesting the geographical origin of goods to which it applies.

3. Classes of Goods

There are 34 classes of goods. Agricultural, horticultural and forestry comes under class 31.

4. Registration of Geographical Indication

The registration of a geographical indication is not compulsory; however, it offers better legal protection to facilitate an action for infringement. The registered proprietor and authorized users can initiate infringement actions. The authorized users can exercise the exclusive right to use the geographical indication.

Applicant: Under the GI Act, any association of persons, producers, organization or authority established by or under the law can be a registered proprietor. The applicant(s) must represent the interest of the producers and their name(s) should be entered in the Register of GI as registered proprietor for the GI applied for.

A producer in relation to goods means the persons dealing with three categories of goods:

(i) If such goods are agricultural goods, produces the goods and includes the person who processes or packages such goods;

(ii) If such goods are natural goods it includes exploiting, trading or dealing;

(iii) If such goods are handicraft or industrial goods includes making, manufacturing, trading or dealing the goods.

It includes any person who trades or deals in such production, exploitation, making or manufacturing, as the case may be, of the goods.

Every application for the registration of a geographical indication should be made in the prescribed form accompanied by the prescribed fee. It must be made in triplicate along with three copies of a Statement of Case accompanied by five additional representations. It should be signed by the applicant or his agent. In case of:

i. an association of persons or producers shall be signed by the authorized signatory;

ii. a body corporate or any organization or any authority established by or under any law for the time being in force shall be signed by the Chief Executive, or the Managing Director or the secretary or other principal officer;

iii. in case of partnership it shall be signed by at least one of the partners.

It must be remembered that the capacity in which an individual signs a document should be stated below his signature. Signatures should be accompanied by the name of the signatory in English or in Hindi and in capital letters.

Every application for registration must contain the following:

a) A statement as to how the GI serves to designate the goods as originating or manufactured from the concerned territory of the country or region or locality in the country, in respect of specific quality, reputation or other characteristics which are due exclusively or essentially to the geographical environment and the inherent natural and human factors that contribute to such quality, reputation or other characteristics.

b) Every application for registration of a G.I. should state the principal place of business in India. A body corporate should state the full name and nationality of the Board of Directors. Foreign applicants and persons having principal place of business, in their home country should furnish an address for service in India. In the case of a body corporate or any organization or authority established by or under any law for the time being in force, the country of incorporation or the nature of registration, if any, as the case may be should be given. The statement contained in the application should also include an affidavit as to how the applicant claim to represent the interest of the association of persons or producers or any organization or authority established under any law

c) The three certified copies of class of goods to which the GI shall apply.

d) The geographical map of the country, region or locality in the country in which the goods originate or are being manufactured. The applicant should also attach documents such as history and reputation of the product.

e) The particulars regarding the appearance of the GI as to whether it is comprised of the words or figurative elements or both.

f) The particulars of special human skill involved or the uniqueness of the geographical environment or other inherent characteristics associated with the geographical indication.

g) A statement containing such particulars of the producers of the concerned good, if any, proposed to be initially registered with the registration of GI.

i) An application to register a geographical indication shall contain a statement of user alongwith an affidavit.

j) Every application of the registration of a geographical indication in respect of any goods shall, on receipt be acknowledged by the Registrar.

k) The acknowledgement shall be by way of return of one of the additional representations with the official number of the application duly entered thereon.

5. Convention Application

To get the priority date of application if a country is a member of convention countries then convention application should contain the following:

1. A certificate by the Registry or competent authority of the Geographical Indications Office of the convention country.

2. The particulars of the geographical indication, the country and the date or dates of filing of the first application.

3. The application must be the applicants' first application in a convention country for the same geographical indications and for all or some of the goods.

4. The application must include a statement indicating the filing date of the foreign application, the convention country where it was filed, the serial number, if available.

6. Register of GI

As per GI Act, the Register of Geographical Indications would be maintained at the Head Office and all the branch offices of Registry, GI wherein specified material particulars would be entered. It also provides for maintenance of the register on the computer or in any other electronic form. The Register shall be kept under the control and management of the Registrar subject to the superintendence and directions of the Central Government. The Register of GI is divided into two parts. Part 'A' consists of particulars relating to registered GIs and part 'B' consists of particulars of registered authorized users. The registration process is similar to both for registration of a GI and an authorized user.

The Appellate Board or the Registrar of GI has the power to remove the GI or an authorized user from the register. The aggrieved person can file an appeal within three months from the date of communication of the order.

Further, in the Act a GI is a public property belonging to the producers or manufacturers of the concerned goods. Thus, there is no provision of any assignment, transmission, licensing, pledge, mortgage or any such other agreement of GI. However, after the death of an authorized user his successor in title can be brought on record.

7. Opposition

Any person within three months from the date of advertisement or re-advertisement of an application for registration or within one month period of extension may file a notice of opposition in writing to the Registrar. The Registrar shall serve a copy of the notice so received on the applicant and if within two months from the receipt by the applicant does not file a counter statement of the grounds on which he relies in support of the application, the application shall be deemed to have been abandoned. If the applicant sends a counter statement, the Registrar shall serve a copy of it to the person giving notice of opposition. Thereafter, the Registrar would dispose of the matter providing opportunity of hearing to the parties and considering the materials on record and the evidences.

8. Term of Registration

The registration of a geographical indication is valid for a period of 10 years. It can be renewed from time to time for further period of 10 years each. If a registered GI is not renewed it is liable to be removed from the register.

In such a case where an identical or nearly similar GI which resembles closely to each other are used by two or more persons being an authorized users, then each of those persons have co-equal rights.

9. Prohibition for Registration of GI

GIs for following are prohibited:

1. If the use of it would be likely to deceive or cause confusion or contrary to any law.
2. If it comprises or contains scandalous or obscene matter or any matter likely to hurt religious susceptibilities of any class or section of citizens of India.
3. If it would otherwise be disentitled to protection in a court.
4. If they are determined to be generic names or indications of goods.
5. If it is, although, literally true as to the territory, region or locality in which the goods originate, but falsely represent to the persons that the goods originate in another territory, region or locality.

10. Benefits of Registration

A registered GI can provide the following benefits:

1. It confers legal protection to facilitate an action for infringement in India.
2. It prevents unauthorized use of a registered GI by others.
3. It boosts export of Indian GIs by providing legal protection.
4. It promotes economic prosperity of producers.
5. It enables seeking legal protection in other WTO member countries.
6. Since GIs are inherently collectively owned, they act as an excellent tool for regional or community-based economic development. Any region that has a speciality associated with it, where a quality link exists or can be established between the product and the region, should consider the

advantages of using a GI to distinguish its product from lower-quality and non-regional competitors.

11. GI v/s Trademark

A GI is like a trade mark, which communicates a message. It tells potential buyers that a product is produced in a particular place and has certain desirable characteristics which are only found in that place e.g. tea producers in other parts of the world may not use the term "Darjeeling" to describe their tea, even if the tea has similar taste. GIs are similar to trade marks in their concept and effect, and can be used to promote national and regional economic development, and are also used strategically in business to promote their products.

A trade mark is a sign which is used in the course of trade and it distinguishes goods or services of one enterprise from the other enterprise while a GI is used to identify goods from a geographical area to which a quality, reputation or other characteristic of a product is essentially attributable.

12. Infringement of Geographical Indication

The registered proprietor or an authorized user of a registered GI can initiate an infringement action. A GI is said to be infringed when:

1. Unauthorized use indicates or suggests that such goods originate in a geographical area other than the true place of origin of such goods in a manner, which misleads the public as to their geographical origin.

2. Use of GI results in unfair competition including passing off in respect of registered GI.

3. The use of another GI results in a false representation to the public that goods originate in a territory in respect of which a GI relates.

It is to be noted that no action can be initiated for relief on the ground of infringement in the case of an unregistered GI. A suitable infringement is to be instituted in the district or higher court having jurisdiction to try the suit. The relief that can be claimed include injunction and damages or account of profits and delivery of infringing labels and indications for destruction or erasure.

Falsifying or falsely applying a GI or selling goods to which GI is applied is an offence. The penalty for the offence could be imprisonment for a term between six months to three years and a fine of fifty thousand to two lakh rupees. The infringing goods can also be forfeited. The court may reduce the punishment under special circumstances.

13. Appellation of Origin

An appellation of origin is a special kind of GI, used on products that have a specific quality which is exclusively or essentially due to the geographical environment in which the products are produced. An appellation of origin is geographical indication, but not all GI are appellation of origin.

14. "Generic" Geographical Indication

If a geographical term is used as the designation of a kind of product, rather than an indication of the place of origin of that product, this term does no longer function as a geographical indication. In case, the product has occurred in a certain country over a substantial period of time, that country may recognize that consumers have come to understand a geographical term that once stood for the origin of the product - for example, "Dijon Mustard," of mustard originally from the French town of Dijon - to denote now a certain kind of mustard, regardless of its place of production.

15. Registered GIs

Few examples of registered GIs are given which have been filed not only from India but from different countries. Darjeeling tea is the first GI registered in October 2004.

Agricultural: Darjeeling Tea (word & logo), Laxman Bhog Mango, Khirsapati (Himsagar) Mango, Fazli Mango grown in the district of Malda (West Bengal), Coorg Orange, Nanjanagud Banana, Hadagali jasmine, Mysore jasmine and Udupi jasmine, Coorg Green Cardamom, Bangalore Blue Grapes (Karnataka), Navara rice, Palakkadan Matta Rice, Malabar Pepper, Alleppey Green Cardamom, Pokkali Rice (Kerala), Allahabad Surkha, Mango Malihabadi Dusseheri, Kalanamak Rice (UP), Eathomozhy Tall Coconut (TN), Naga Mircha, Naga Tree Tomato (Nagaland), Mahabaleshwar Strawberry, Nagpur Orange (Maharashtra), Bhalia Wheat (Gujarat), Ganjam Kewda Flower (Odhisha), Arunachal Orange (Arunachal Pradesh), Sikkim Large Cardamom (Sikkim), Mizo Chilli (Mizoram), Tripura Queen Pineapple (Tripura), Tezpur Litchi (Assam), Khasi Mandarin, Memong Narang (Meghalaya), Kachai Lemon (Manipur),

Handicrafts and Manufactured: Mysore Agarbathi (Karnataka), Mysore Sandalwood Oil and Mysore sandal soap (Karnataka), E.I. Leather (TN), Meerut Scissors (UP). Aranmula Kannadi (Kerala), Pochampalli Ikat (AP), Chanderi Fabric (MP), Mysore Silk (Karnataka), Kancheepuram Silk, Thanjavur Paintings, Temple Jewellery of Nagercoil (TN), Kullu Shawl, Kilu Rumal (HP), Orissa Ikat, Konark stone carving (Odisha), Silver Filigree of Karimnagar (Telengana), Muga Silk, Muga Silk of Assam (Logo) (Assam), Bastar Dhokra, Bastar Wooden Craft, Bastar Iron Craft (Chattisgarh), Kathputlis of Rajasthan, Blue Pottery of Jaipur, Makrana Marble (Rajasthan), Sankheda Furniture, Kutch Embroidery, Patan Patola (Gujarat), Kani Shawl, Kashmir Pashmina, Kashmir Walnut Wood Carving (J&K), Lucknow Chikan Craft, Banaras Brocades and Sarees (Logo), Lucknow Zardozi, Moradabad Metal Craft (UP),

Phulkari (Punjab, Haryana & Rajasthan), Bhagalpur Silk (Bihar), Shaphee Lanphee, Wangkhei Phee, Moirang Phee (Manipur).

Food stuffs: Bikaneri Bhujia (Rajasthan).

Spirits: Feni (Goa), Nashik Valley Wine (Maharashtra)

Goods of Other countries: Peruvian Pisco (Peru), Champagne, Cognac (France), Napa Valley (USA), Scotch Whisky (UK), Prosciutto di Parma (Italy), Porto, Douro (Portugal), Tequila (Mexico).

16. Protection of Appellations of Origin and International Registration

The Lisbon Agreement, concluded in 1958, was revised at Stockholm in 1967, and amended in 1979. The Lisbon Agreement provides for the Protection of Appellations of Origin. The Lisbon Agreement created a Union which has an Assembly. Every State member of the Union that has adhered to at least the administrative and final clauses of the Stockholm Act is a member of the Assembly. Champagne, Cognac, Roquefort, Chianti, Porto, Tequila, Darjeeling, etc. are some examples of names which are associated with products of a certain nature, quality and geographical origin. Geographical indications are protected in accordance with international treaties and national laws under a wide range of concepts, including sui generis laws for the protection of geographical indications or appellations of origin, trademark laws in the form of collective marks or certification marks, laws against unfair competition, consumer protection laws, or specific laws or decrees that recognize individual geographical indications. Securing protection for such indications in other countries became complicated due to differences in legal concepts existing from country to country in this regard. As its name indicates, The Lisbon Agreement for the Protection of Appellations of Origin and Their International Registration (hereinafter referred to as 'the Lisbon Agreement') was specifically concluded in response to the need for an international system that would facilitate the protection of a special category of such geographical indications, i.e. "appellations of origin", in countries other than the country of origin, by means of their registration with WIPO through a single procedure, for a minimum of formalities and expenses.

An appellation of origin is a special kind of geographical indication. It generally consists of a geographical name or a traditional designation used on products which have a specific quality or characteristics that are essentially due to the geographical environment in which they are produced. Under Article 2 of the Lisbon Agreement, appellations of origin are defined as follows: 'Appellation of origin means the geographical denomination of a country, region or locality which serves to designate a product originating therein, the quality or characteristics of which are due exclusively or essentially to the geographical environment, including natural and human factors. The country of origin is the country whose name, or the country in which is situated the region or locality whose name, constitutes the appellation of origin that has given the product its reputation. In order to qualify for international registration, the protection of the appellation of origin must have been formalized first in the country of origin, either by means of legislative provisions, or administrative provisions, or a judicial decision or any form of registration.

Appellations of origin are a collective tool for producers to promote the products of their territory and also preserve their quality and reputation acquired over time. The use of the protected appellation of origin is reserved to those producers that are able to meet a number of specifications, including geographical area of production, methods of production, product specificities, etc. When the name of a product receives protection as an appellation of origin, the local communities benefit from the positive impact, in various ways: the appellation increases production and creates jobs (differentiation strategy which results in higher prices while also helping to sustain the production of traditional products) the appellation enables improved redistribution of the added value across the whole production chain, from the producer of the raw material to the manufacturer.

Once protected in the country of origin, the holders of the right to use the appellation of origin may request their Government to file an application for international registration under the Lisbon Agreement. International registration of an appellation of origin takes place at the request of the 'country of origin', in the name of interested parties (i.e. any natural person or legal entity, public or private, having, according to their national legislation, a right to use such appellation). The International Bureau then notifies the Competent Offices of the other Contracting Parties to the Lisbon Agreement of any new international registration of an appellation of origin. International registration is subject to payment of a single 500 Swiss francs fee. The international registration of an appellation of origin ensures the protection of that appellation without renewal, for as long as it is protected in the country of origin. A registered appellation may not be presumed to have become generic in a Contracting State as long as it continues to be protected in the country of origin. In addition, the other Contracting States are under the obligation to provide a means of defense against any usurpation or imitation of an internationally protected appellation of origin in their territory. In principle, an internationally registered appellation of origin must be protected in all countries of the Lisbon system. However, these countries do have the right to refuse such protection, for example, on the ground that, in their territory, the appellation of origin corresponds to a protected trademark or to a generic indication of a particular product. They can do so by notifying a declaration of refusal to WIPO within one year from the receipt of the notification of registration issued by the International Bureau. When a refusal has been initially issued, but it appears over time that the conditions that have motivated such refusal are no longer valid, a country may either issue a withdrawal of refusal or a statement of grant of protection. If no refusal is submitted, the appellation of origin will be considered automatically protected for as long as it is registered (unless a court in the country invalidates the effects of the registration in the country in question).

Selected reading

Chawla, H.S. and Singh, A.K. 2005. Intellectual Property Rights. Vol II: Copyrights, Trade Marks, Trade Secrets and Geographical Indications. Pantnagar University Press, pp. 75.

http://www.wipo.int/treaties/en/

James, T.C. 2010. Protection of geographical indications and socio-economic development. In: Compendium on Intellectual Property Rights and Development: A national Perspective (Eds Yeshodharan, E.P. et al.), Kerala State CST&E, Thiruvananthapuram, Kerala, pp. 65-70.

The Geographical Indications of Goods (Registration and Protection) Act, 1999 along with The Geographical Indications of Goods (Registration and Protection) Rules 2004, Universal Law Publishing Co., Delhi, 2011.

Trade Secret

Knowhow is another important form of intellectual property generated by R&D institutions that do not have the benefit of patent or copyright protection. Such know-how is kept undisclosed as trade secrets. A Trade Secret or undisclosed information is any information that has been intentionally treated as secret and is capable of commercial application with an economic interest. It protects information that confers a competitive advantage to those who possess such information, provided such information is not readily available with or discernible by the competitors. A trade secret is any information that gives a company a competitive edge over competitors and which the company maintains as secret and away from public knowledge. Trade secrets often include private proprietary information. Examples of trade secrets are the Coca Cola company brand syrup formula, Polaroid company instant film chemical formula, etc. The nature or the identity of a product is maintained secret for as long as the company can keep this information from becoming public knowledge.

Trade secret is commonly defined as any sort of information, formula, pattern, device or compilation of information which is used in one's business and which gives human opportunity to obtain an advantage over competitors who do not know or use it. The information can be any formula for chemical compound, a device/machine, pattern, device or compilation of information, process of manufacturing, treating or preserving materials, a pattern for a machine or other device, a computer program or a list of customers, financial standing of business, employee records, credit rating of customers, production information, business methods, test data, research & technical reports, process manuals, source code, data file structure, design, drawings etc. It is a legal term for confidential business information. This information allows a company to compete effectively. The holder of trade secret takes measure to preserve its confidential nature.

Broadly speaking, a trade secret must follow the three basic criteria:
1. The information must not be "generally known or readily ascertainable" through proper means to the public.

2. The information must have "independent economic value due to its secrecy."

3. The trade secret holder must use "reasonable measures under the circumstances to protect" the secrecy of the information.

Trade secret remains confidential for indefinite period of time as per the will of the proprietor provided the security and its confidentiality is not breached.

Trade secrets are embodied in undisclosed information. If any improper method is used to obtain a competitor's trade secret it is an infringement and is subject to injunctions and damages. However, if a trade secret becomes public knowledge by independent discovery or other means, then it is not punishable by court.

1. Indian Legal Framework

There is no specific legislation regulating the protection of trade secrets in India. Indian courts have upheld trade secret protection on the basis of principles of equity. India follows common law approach of protection and all matters relating to it are generally covered under the Indian Contract Act, 1872. The provision relating to restraint of trade secret agreement is under Section 27. Civil laws concerning employment and general contracts, criminal and commercial law provisions cover the protection under trade secrets. So, if the information constituting trade secret is leaked, legal action can be brought against the parties who have leaked it under the Law of Contracts. However, in such a case the protection of trade secret will be lost and it becomes available in public domain. In case of trade secrets there is no concept of registration or term but protection is obtained through legal (non-disclosure) agreement between the owner of the secret and those who have been given access to it. Law prevents wrongful taking of confidential or secret information. Independent development and reverse engineering by another party are defenses to claims of trade secret theft.

Trade secret protection is a state right. A majority of countries have adopted the Uniform Trade Secret Act. Other countries have a law or laws similar to the Uniform Trade Secret Act, although, the particular country should be checked before reliance is placed on trade secrets in a foreign country. The rights will seem to be national or international in scope, since registration of trade secrets is not required.

2. Trade Secret Under TRIPs Agreement

Paris convention and Trade Related Aspect of Intellectual Property Rights (TRIPs) Agreement are two major documents in IP management and rights enforcement. Article 39.1 of the TRIPs Agreements provide that "in course of ensuring effective protection against unfair competition as provided in Article 10bis of the Paris Convention, member shall protect "undisclosed information" of the sort which is described in paragraphs 2 and 3 of Article 39. Paragraph 2 describes the general category of confidential information, which is protected in common law countries through judge-made law, rather through statute. Article 10 bis (2) defines as an act of unfair competition "any act of competition contrary to honest practices in industrial and commercial practices". Article 10bis lists three particular practices which are to be prohibited. The TRIPs agreement negotiators were anxious to preserve the confidentially of test data submitted to government approval agencies. Given the long approval process, particularly for pharmaceutical products, the opportunity for wrongful appropriation of such data by competitors was self evident. This concern is accommodated by Article 39.3 which provides that:

Members, requiring as a condition of approving the marketing of pharmaceutical or of agricultural chemical products which utilize new chemical entries, the submission of undisclosed test or other data, the origination which involves a considerable effort, shall protect such data against unfair commercial use. In addition, Members shall protect such data against disclosure except where necessary against unfair commercial use.

It should be noted that article 39.3 contains three limitations. First, it applies only to pharmaceutical products and chemical agricultural products; secondly the protection is extended only against unfair competition uses; and thirdly the government authority is exempted from the requirement of confidentially in the public interest. Thus, it has been held that a government accrediting agency may use the confidential test data of an applicant when considering applications by other applicant in respect of similar products.

Where, the specifications of an invention relatable to an article substance covered under sub-section (2) of section 5 have been recorded in a document or the invention has been tried or used, or, the article or the substance has been sold, by a person, before a claim for a patent of that invention is made in India or a convention country, then, the sale or distribution of the article or substance by such person, after the claim referred to above is made, shall be deemed to be an infringement of exclusive right to sell or distribute under sub-section (1):

Provided nothing in this sub-section shall apply in a case where a person makes or uses an article or a substances with a view to sell or distribute the same, the details of invention relatable thereto was given by a person who was holding an exclusive right to sell or distribute the article or substance.

3. Trade Secret v/s Patent

In a patent, the protection afforded is temporary and it ceases on the expiry of the patent. Further, since the information must be revealed while filing the patent, it becomes publicly known. In contrast, a properly maintained trade secret can remain a secret for a much longer period of time. A company can still protect the information even if the information is or might become generally known or readily ascertainable. Patents, copyright, and trade marks laws will provide protection for certain information even when the information is generally known or readily ascertainable. Contracts can also provide rights that exceed the bounds of trade secret law. A decision to pursue patent protection instead of trade secret protection is an involved decision. Such a decision, however, turns in part on your likelihood of succeeding on showing that the information "is not generally known or readily ascertainable."

4.   Trade Secret v/s Copyright

Copyright is given for an expression on idea. However an idea may attract legal protection as confidential information. The requirement of copyright law is to show specific expression of the idea in scenarios or scripts, in writing or other recorded form. However, this is not necessary in the law of confidence. It is enough that the content of the idea is clearly identifiable, original, potential of commercial attractiveness and capable of being realized in actuality. The technical information does not have to be novel or attain any level of inventiveness.

5.   Protecting Trade Secrets

Trade secret by definition is not disclosed to the world at large. Instead, owners of TS keep their special knowledge out of the hands of competitors through a variety of civil and commercial means such as Non Disclosure Agreements (NDAs) and Non Compete Clauses (NCCs). It is important to bear in mind that a TS need not be something that is novel nor should it have any real or intrinsic value to be protected. The only requirement is that it must be a secret and therefore to be protected against misappropriation, sabotage and piracy.

Trade secret protection is a must for virtually any business. It's most often not addressed until an employee or competitor obtains and uses against your valuable secret information, thereby stealing company's sales, customers, technology base, damaging financial information, or others.

The holder of trade secret should take proper measures to preserve its confidential nature. Some examples to illustrate this are:

1. The employees are bound to maintain confidentiality of the information through some kind of confidentiality agreements.

2. In case of technology transfer, the licensee and the licensor may be bound by a confidentiality agreement for not disclosing the business details to their competitors.

3. In joint research programmes confidentiality agreements may ensure that the results of the research are not made known to any outsider and if there is a need to do so, it will be with mutual consent of the collaborating partners.

Therefore the trading community and corporate world must devise multiple strategies to protect their trade secret.

5.1 Employment Agreement

If a company fails to protect the proprietary information it will allow the competitors and ex-employees to reduce one's profits. The trade secret laws will help prevent such misfortune if the company acts in accordance with its requirements. For example if a company's top employee leaves the company, since he knows every minor area of the company and then that employee can either sets up his/her own business in direct competition with you or become an employee of your toughest competitor. One can stop this individual if the company had protected its business information properly under the trade secret laws. Proper protection requires action today to be ready for tomorrow.

Depending upon their needs, business should include suitable confidentiality, NDAs and NCCs in agreement with employees. These may include the type of information that is likely to be disclosed, the manner in which it should be used and restrictions on disclosure post-termination.

5.2 Trade Secret Policy

Trade secret policy is a must for businesses that heavily rely on their TS. A basic step is to develop such a policy to identify and prioritize the business secrets based on their value and sensitivity. Employees must be informed about such policy and consequences of its breach. Employees must agree to abide by the policy and sign an acknowledgement to that effect.

5.3 Non-Disclosure Agreements

Trade secret rights are mainly kept and enforced through agreements between employers and employees. These are non-patented. Usually at the time employment begins, an employer makes an employee sign an agreement that grants the employer trade secret protection. The agreements protect the company by preventing its competitors from enticing key personnel since these individuals cannot divulge the trade secret material without incurring severe penalties.

Non-disclosure agreements should be signed before a person see, hear or otherwise learns confidential information of the company. In case of a new employee, the agreement should be signed before the first day of work. If they sign after they start working or otherwise obtain access to the information, they should be given something of value in exchange for their signature (this may be necessary to make the contract binding). Non-disclosure agreements are not used with just employees, but rather anyone who has access to the confidential information. A distinction has to be made between agreements, which have been read, and those, which have not. An agreement that has not been read may provide contract rights, but does not provide significant notice value. Signed contracts, along with audits, are perhaps the most powerful tools in trade secret law. Beyond notice, contracts can be used to broaden the trade secret holder's rights and provide basis for asserting misappropriation (i.e., breach of contract). A well-prepared non-disclosure/non-compete contract is a must. Court's seem to decide for trade secret holders in the presence of a contract and against the holder in the absence of a contract. Criminal prosecution of an employee who steals trade secrets from their employer is a recognized remedy. Some examples of trade secret misappropriation are given in Boxes 1 to 3.

5.4 Confidentiality Measures

Measures that a company can take are often considered under both notice measures and physical security measures. Notice measures are those measures that put persons who come in contact with the information on notice that the information is to remain secret. Physical security measures are those measures that prevent people who do not need-to-know the information from coming in contact with the information (e.g., confidentiality barriers).

An employee is responsible for keeping information confidential till the employer had not expressed a desire to keep the information confidential. All those with access to confidential information must be given notice as to what information is to remain confidential. A good use of notice measures involves frequent and clear instructions on confidentiality.

Any communication which identifies that the information is confidential or how to handle confidential information will work as a notice measure. Employee handbooks, newsletters, and signs are common examples. The number and types of procedures are limitless. The motive is to express which information is to be confidential, either orally, in writing, through actions, and any other method of communication.

5.5 Security Systems

Access to TS and confidential information may be restricted to only selected personnel who have to undergo proper security checks.

The location of information should be separated into different areas (e.g., file cabinets, rooms or buildings). Only people that have a need-to-know the confidential information should have access to the information in that area. Signs stating "employees only", "authorized personnel only", "restricted access", "private" or similar phrases will help in keeping people away from getting into these restricted access areas.

In case of an electronic environment, the businesses should use adequate software programmes, virus scans, firewalls and other security and authentication technologies to safeguard their TS.

The other methods of direct control of information are:

1. Passwords or a separate computer system. A computer system may give off a warning when someone tries to break the password.

2. Confidential information should be put away when not in use. A blanket may be thrown over a machine. A drawer may be closed and locked.

3. The disposal of confidential information should be carefully handled.

Machines built according to trade secret knowledge should be disassembled.

A visitor does not necessarily know about the procedures of a company. Keeping a logbook at reception suggests that visitor will be monitored and can be checked. They can also be refused without having an appointment. A time schedule for visitors can be made which allows the employees to be prepared for the visit. Visitors should be escorted at all times and they should be kept off from the confidential information. Visitors can be provided batches to distinguish them from employees.

Some of the security measures for protecting trade secrets are given below:

1. Transform information into tangible form as evidence of existence.
2. Maintain record of the movement of information within and outside the company.
3. Maintain date of creation of information
4. Notify the recipient of trade secret in writing
5. Establish security coded ingredients or data
6. Have separate departments of the company so that each department has a particular type of information but not the whole.
7. Stamp documents as confidential or secret
8. Enter into vendor secrecy agreement
9. Hold exit interviews to obtain return of company documents
10. Educate employees about existence of confidential information and need to keep it secret
11. Sales and marketing staff should be trained in not leaking the plans of the company
12. Monitor internet traffic to ensure that keywords or specific phrases are not transmitted.

5.6 Comprehensive Documentation

It is important for businesses to keep a track of the trade secrets that are developed and have sufficient records to show that TS was developed by them and belonged to them. These records would act as evidence in case of disputes. It would also be useful for such businesses to conduct a TS audit at regular intervals to update any changes. Besides record keeping is must and following things should be kept in mind.

- Permanently bound record book.
- Loose sheets reduce credibility.
- Sequentially numbered notebook pages

- Good quality paper & light stable ink to last 20-25 years.
- Legible, correct and duly signed entries.
- Describe all experimental details, setups
- Carefully record all observations. Affix loose sheets, drawings, graphs and make references. Avoid incomplete pages, draw lines through unused pages.
- Record all novel concepts and ideas avoiding opinions like 'the idea is not novel' or 'unsuccessful experiment.
- Simple language for understanding of judges.

Since there is no documentary evidence that the trade secret was originally created by the proprietor, it is essential to maintain proof of creation of trade secret either by mailing the information to oneself and retaining postmarked and sealed envelope or by depositing a copy of the information with a third party that would maintain a dated copy.

6. Case Studies

Misappropriation is wrongful taking of trade secret information. A wrongful taking can occur in a variety of manners. It includes breach of contract, breach of judiciary obligation, theft, or other legal wrongful deeds. Some case studies are given in Boxes 1-3.

Box 1
Trade secret misappropriation of Brandon Process
This is an example of trade secret litigation between two pharmaceutical companies and at stake were billions of dollars and thousands of jobs. On one side was Wyeth – one of the world's foremost manufacturers and makers of Premarin, a hormone replacement therapy drug which has been sold in US since 1942. Wyeth makes Premarin from the estrogens found in the pregnant mare urine (PMU) using a closely guarded secret chemical process known as Brandon Process named after the manufacturing facility in Brandon, Manitoba. Brandon Process extracts estrogens from PMU and transforms them in to dry powder – Preserved condensed urine desiccated (PCUD). Brandon Process is neither generally known nor readily ascertainable. For more than sixty years, no company has ever been able to replicate the Brandon Process. Brandon Process is not patented. In 1940's and 1950's Wyeth obtained patents on several methods discovered in connection with its estrogen extraction research and the last patent expired in 1975. On the other side was Natural Biologics – David Saveraid (Agricultural salesman) is the founder and President, a start up company in Albert Lea, Minnesota, where it extract natural conjugated estrogens from PMU in the form of a dry bulk product. It intended to market a generic form of Premarin. An eight year court battle fought over trade secret misappropriation. Finally the Court concluded that measures taken by Wyeth's information protection efforts fell in the upper end of the range of reasonableness for corporate security programs. Natural Biologics was "a company built upon misappropriation". Judge found that the Company's CEO

(David Saveraid (Agricultural salesman) had conspired with a former Wyeth scientist (Dr. Irvine, a chemist) to steal the company's secret Brandon Process. Judge ordered natural Biologics shut down, ordered the destruction of all research materials and drug supplies at the company and banned its leaders from ever seeking to re-enter the estrogen replacement market.

Box 2

Imprisonment to Ex-Dupont Chemist for Trade secret misappropriation

Gary Min also known as Yonggang Min, a former chemist of DuPont Co., was sentenced to 18 months imprisonment and a fine of $30,000, after he pleaded guilty. Min was accused of stealing $400 million worth of confidential DuPont technology to take overseas. Judge Sue Robinson termed the sentence as 'fair', and justified it by stating that there was no evidence of any master plot and that Mr. Min's remorse and contrition are beyond question. Moreover, the actual harm to DuPont was found to be negligible. This incident of trade secrets theft surfaced only when Mr. Min started to look for possible job opportunities in July 2005.

Box 3

Jury Awards compensation and damages against Sears for trade secret misappropriation

The U.S. District Court for the Northern District of Illinois handed down the largest over award of compensation and punitive damages in the case of trade secrets. The dispute arose when Mr Bob Kopras and its company Roto Zip Tool Corp, complained against Sears, Roebuck and Co. for misappropriation of trade secrets. The dispute dates back almost a decade when Roto Zip was one of the major suppliers of Sears. Mr. Kopras, in the year 1999 had disclosed, confidentially, to Sears the draft of a new model of hand held combination saw. No contract was signed between the parties as they failed to agree on the price. With time Mr. Kopras developed on his draft, unknown to the facts that Sears have employed a Chinese company to make a copy of the design against the non disclosure agreement with Mr. Kopras. Kopras launched his finished product, and after a week, in August 2001, Sears introduced the Craftsman Combination Tool. Kopras filed a suit in 2004 complaining of misappropriation of trade secrets. The jury noted that Sears have been selling the device, and that there is a dramatical dip in the sales of Roto Zip's rotary saw. The jury found Sears guilty of willful and malicious trade secret misappropriation and breach of contract.

Selected reading

Chawla, H.S. and Singh, A.K. 2005. Intellectual Property Rights. Vol II: Copyrights, Trade Marks, Trade Secrets and Geographical Indications. Pantnagar University Press, pp. 75.

Nomani, Md. Zafar Mahfooz. 2012. Synergy of trade secret legislations, competition laws and innovation: Protection statutes under emerging intellectual property rights regime: A comparative and Indo-US perspective. *In*: Protecting intellectual property in life sciences (Eds. Keating, D., Agnihotri, A. and Verma A.), Amity University Press, NOIDA, India

Semiconductor Integrated Circuits Layout-Design

Integrated circuits date back to the 1950's when Geoffrey W.A. Dummer, a British radar scientist in the Defense Ministry made a public symposium on the topic. He failed to assemble the IC in 1956 but J. Kilby and R. Noyce succeeded to build the IC's in 1950. Everyday life is becoming extremely dependant on machines, communication, transport, manufacturing, and medical technology. All these are dependent on the existence of IC's. Almost all equipment in the world use IC'. Examples include phones, watches, TVs, computers, electronic doors, cars, medical scanners and many more.

IC's are creations of the human mind just like other forms of IP. A lot of capital commitment, research & development, time, human resources and coordination are needed to come up with IC creations. IC's can be copied just by making replicas of each layer on the IC. For this reason it became necessary to protect the topography (layout design) of IC's to protect the intellect and the resources that would have been spent in coming up with such a component. They are a source of competitive advantage. In the Computer industry IC's are becoming smaller and more efficient, through high level R&D schemes. IBM and Intel are major spenders of capital on the IC's R&D.

What is protected here is the topography (layout design) which is a three dimensional representation of the layers of an Integrated Circuit, showing interconnections of both the active and passive elements embedded within the IC. The representation must be in such a way that it reflects a flow used for manufacturing.

Thus, it provides protection for semiconductor IC layout designs. The layout-design of a semiconductor integrated circuit means a layout of transistors and other circuitry elements and includes lead wires connecting such elements and expressed in any manner in semiconductor integrated circuits. Semiconductor Integrated Circuit means a product having transistors and other circuitry elements, which are inseparably formed on a semiconductor material or an insulating material or inside the semiconductor material and designed to perform an electronic circuitry function.

The layout of transistors on the semiconductor integrated circuit or topography of transistors on the integrated circuit determines the size of the integrated circuit as well as its processing power. That is why the layout design of transistors constitutes such an important and unique form of intellectual property fundamentally different from other forms of intellectual property like copyrights, patents, trademarks and industrial designs.

1. International Treaties and Conventions

1.1 Washington Treaty on Intellectual Property in Respect of Integrated Circuits

The Washington Treaty was adopted in 1989 and provides protection for the layout designs (topographies) of integrated circuits. Article 35 of the TRIPs agreement gives members an obligation to protect IC topography (layout-designs). The Treaty has not yet entered into force, but has been ratified or acceded to by the some States: Bosnia and Herzegovina, Egypt and Saint Lucia.

2. Semiconductor Integrated Circuits Layout-Design Law in India

The Semiconductor Integrated Circuits Layout-Design (SICLD) Act, 2000 and The Semiconductor Integrated Circuits Layout-Design Rules, 2001 was passed to fulfill India's obligations as a TRIPs signatory. It provides protection for Semiconductor Integrated Circuits Layout Designs. The main purpose of the Act is to provide for routes and mechanism for protection of IPR in Chip Layout Designs created and matters related to it.

The Act is implemented by the Department of Information Technology, Ministry of Information Technology. The Act is applicable for IC Layout-Design IPR applications filed at the Registry in India. The Semiconductor Integrated Circuits Layout-Design Registry (SICLDR) is the office where the applications on Layout-Designs of integrated circuits are filed for registration of created IPR. The Registry has jurisdiction all over India.

The important provisions of the Act are:

- Jurisdiction to the whole of India;
- SICLD Registry – where the layout-designs of integrated circuit chips can be registered;
- Defines layout-designs of integrated circuits which can be registered under the Act;
- Duration of registration of layout-designs;
- Rights conferred by registration;
- Infringement of layout-designs;
- Procedure for assignment and transmission of registered layout-design;
- Appellate Board as a forum of redressal;
- Treatment of Royalties;

- Provisions in case of national emergency or extreme public urgency;
- Penalties;
- Provision for agents;
- Reciprocity provision with other recognized countries.

3.   Criteria for Registration of Chip Layout Design

An IC layout design is registrable under Sec. 7(1) if it is:

a) Original

b) Not commercially exploited anywhere in India or in a convention country (not commercially exploited for more than 2 years from the date on which an application for its registration has been filed either in India or in a convention country)

c) Inherently distinctive

d) Inherently capable of being distinguishable from any other registered layout design

Under Sec. 7(2) a layout-design shall be considered to be original if it is the result of creator's own intellectual efforts and is not commonly known to the creators of layout designs and manufacturers of semiconductor integrated circuits at the time of creation. Further, a layout-design consisting of such combination of elements and interconnections that are commonly known among creators of layout-designs and manufacturers of semiconductor integrated circuits shall be considered as original if such combination taken as a whole is the result of its creators own intellectual efforts.

Under Sec. 7(3) it has been made clear if an original layout-design has been created in execution of a commission or a contract of employment, shall belong to the person who commissioned the work or to the employer, if there is no contractual provision to the contrary.

3.1   Prohibition from Registration

A layout-design is prohibited from registration under Sec. 7 if:

i. It is not original.

ii. It has been commercially exploited in India or in a convention country.

iii. It is not inherently distinctive. Also, those which are not capable of being distinguishable from any other registered layout-design cannot be registered.

4.   Person Entitled to Protection of Layout-Designs

Under Sec. 8 any person(s) who:

1) is a creator of a layout design and desires to register it;

2) is an Indian national or national of country outside India which accords to citizens of India similar privileges as granted to its own citizens in respect of registration and protection of layout-designs and;

3) has principal place of business in India or if he does not carry out business in India, has place of service in India. In case of joint application the principal place of business in India of the applicant whose name is first mentioned in the application.

5.   Steps for Registration of a Layout-design

1) Filing of application by the creator of the layout-design at the SICLD Registry.

2) The Registrar may accept, refuse the application or accept with some modifications.

3) The accepted applications shall be advertised within 14 days of acceptance.

4) Any opposition to the advertisement can be filed within 3 months from the date of advertisement.

5) The counter-statement to the notice of opposition, if any, to be filed within 2 months from the date of receipt of copy of notice of opposition from the Registrar.

6) A copy of the counter statement provided to the opposing party.

7) The Registrar may take hearing with the parties.

8) The Registrar will decide on the originality of the layout-design and grant or reject the application for registration based on the conclusions reached by him.

9) Aggrieved party can appeal to Appellate Board or in its absence Civil Court for relief on any ruling of the Registrar.

6.   Documents to be Submitted Along with Application

1) 3 sets of 2D/3D drawings which describe the layout-design applied for registration and or 3 sets of photographs of masks used for the fabrication of the semiconductor integrated circuit by using of the layout-design applied for registration, and or drawings which describe the pattern of such masks.

2) Semiconductor integrated circuit [where an integrated circuit has been made using layout-design applied for registration].

3) Any related information sought by Registry/Registrar.

In case the applicant for registration of layout-design makes a request in writing for maintaining the secrecy of the layout-design, he may attach in place of the drawings or photographs, the three sets of partially blocked drawings or photographs of such layout-design to the satisfaction of the Registrar. The Registrar may inspect the complete drawing or photographs of such layout-design. The blocking of such drawing or photograph shall be to the satisfaction of the Registrar such that it does not hamper the identification of the applied-for layout-design. The blocked out area of such drawing or photograph should not be greater than the area of the remaining portion of layout-design.

7. Opposition

Any person can give notice of opposition to the registration in writing in the prescribed manner to the Registrar within three months from the date of the advertisement or readvertisement of an application for registration.

8. Duration of Registration

The registration shall be only for a period of 10 years. A period of 10 years is counted from the date of filing an application for registration or from the date of first commercial exploitation anywhere in India or in any convention country or country specified by Government of India whichever is earlier (Sec. 15).

9. Rights

The registered proprietor of layout-design shall have the exclusive rights to use the lay-out design and to obtain relief in respect of infringement.

10. Infringement and Penalties

Under Sec. 18 a layout-design is infringed by a person who, not being the registered proprietor of the layout-design or a registered user thereof:

a) does any act of reproducing, whether by incorporating in a semiconductor integrated circuit or otherwise in its entirety or part thereof, except an act of reproducing the whole or part is not original as per prohibition of registration of layout designs under Sec. 7(2)

b) does any act of importing or selling or otherwise distributing for commercial purposes a registered layout-design or a semiconductor integrated circuit.

However if such act is performed for the limited purposes of scientific evaluation, analysis, research or teaching shall not constitute act of infringement.

10.1 Penalties for infringement

Any person who contravenes knowingly and willfully any of the provisions of section 18 as mentioned above shall be punishable with imprisonment for a term which may extend to three years or with fine which shall not be less than fifty thousand rupees but which may extend to ten lakh rupees or with both.

Any person who makes any representation with respect to layout-design not being a registered layout-design, to the effect that it is a registered layout-design shall be punishable with imprisonment upto six months or with fine upto fifty thousand rupees or with both.

Any person who uses on his place of business, or on any document issued by him, or otherwise, words which would reasonably lead to belief that his place of business is, or is officially connected with the Semiconductor Integrated Circuits Layout-Design Registry shall be punishable with imprisonment upto six months or with fine or with both.

If any person makes, or causes to be made a false entry in the register shall be punishable with imprisonment upto two years or with fine or with both.

11. Assignment and Transmission

The person for the time being included in the register as proprietor of a layout-design has power to assign the layout design, and to give effectual receipts for any consideration for such assignment (Sec. 20). Notwithstanding anything in any other law to the contrary, a registered layout-design be assignable and transmissible with or without the goodwill of the business concerned.

When an assignment of a registered layout-design is made otherwise than in connection with the goodwill of business, the assignment shall not take effect unless the assignee, not later than the expiration of six months from the date on which the assignment is made or within such extended period, if any, not exceeding three months in aggregate, as the Registrar may allow with certain conditions (Sec. 22).

Where a person becomes entitled by assignment or transmission to a registered layout-design, he shall apply in the prescribed manner to the Registrar to register his title. In cases where the validity of an assignment or transmission is in dispute between the parties, the Registrar may refuse to register until the rights of the party have been determined by a competent court (Sec. 23).

12. Registered Users

Under Sec. 24 any person other than the registered proprietor of a layout-design can be registered as a registered user thereof. It is proposed that a person should be registered as a registered user of a layout-design for which registered proprietor and the proposed reregistered user shall jointly apply in writing to the Registrar (Sec. 25).

13. Rectification and Correction

Any person aggrieved by the absence or omission from the register of any entry or by any entry in the register without sufficient cause or wrongly remaining on the register may apply in the prescribed manner to the Appellate Board or the Registrar. The Appellate Board or the Registrar as the case may be, may make such order for making or expunging or varying the entry as it may think fit. This shall be done after giving notice in the prescribed manner to the concerned parties (Sec. 30). Likewise The Registrar may correct any error in the name, address or description of the registered proprietor of a layout-design, enter any change in the name, address or description of the person registered as proprietor and also registered user following the procedure made in the rules (Sec. 31).

14. Appellate Board

Under section 32 of the Act, a Layout-Design Appellate Board is to be established. Till the establishment of the Appellate Board, The Intellectual Property Appellate Board established under Sec. 83 of the Trade Marks Act, 1999 shall exercise the jurisdiction, powers and authority conferred on the Appellate Board. The appellate board consists of Chairman, technical members and other staff. The technical member shall be appointed under this Act for constituting the bench. It has been established for making appeals to the decisions made by the Registrar. Appeal should be made within three months from the date of the decision, order or direction of the Registrar.

Selected reading

http://www.dcip.gov.zw/
http://www.wipo.int/treaties/en/
The Semiconductor Integrated Circuits Layout-Design Act, 2000 along with The Semiconductor Integrated Circuits Layout-Design Rules, 2001, (Universal Law Publishing Co., Delhi), 2010.

CHAPTER 10

Technology Management: Licensing and Commercialization

When an inventor creates intellectual property, inventor initially uses the same for its commercial exploitation to gain financial benefits out of it. But once its value has been established or created, inventor begins to valorize IP, i.e. increase its value through licensing, technology transfer, etc. IP valorization is a tool of generating wealth through intellectual property. The valorization of capital is a theoretical concept created by Karl Marx in his critique of political economy. The meaning of **valorization** is a use or application of something (an object, process or activity) so that it makes money or generates value. Thus something is valorized if it has yielded its value. Related terms are exploitation, commercialization and value extraction. IP valorization can therefore be understood as the process of making use of intellectual property.

Intellectual property assets can be commercially exploited by their owner or with the permission of the owner by others. One way for others to exploit intellectual property is through 'licensing' the intellectual property from the owner. Licensing, the right granted by an owner of an asset to another to use that asset while continuing to retain ownership of that asset, is an important way of creating value with these assets. Licensing creates an income source, disseminates the technology to a wider group of users and potential developers and acts as a catalyst for further development and commercialization. The word "license" simply means permission granted by the owner of the intellectual property rights to another to use it on agreed terms and conditions, for a defined purpose, in a defined territory and for an agreed period of time. Licensing of intellectual property is often considered in three broad categories, namely technology licenses, publishing and entertainment licenses, and trademark and merchandising licenses. Licensing of inventions related to biotechnology come under technology licenses.

Technology means many things to many people. A popular definition of technology is that "technology is the practical use of scientific information." Therefore, broadly speaking, technology refers to end products of scientific research and development in the form of inventions and know-how which are used as tools or processes for creating new or improved products and services that better serve the needs of the market. There is often a tendency to equate one patent with one technology. This is rarely the case nowadays. Increasingly, a number of patents together are responsible for a technology and a number of technologies for a product, for example, a camera or a car. Such technology may be acquired either through R&D undertaken by the company itself, in cooperation with others, or by acquiring technology developed by others which may be on offer in the market.

Given the intangible character of technology, its use by one does not detract its use by another. In other words, it can be used simultaneously by many users for the same or different purposes without impacting in any way on its quality or functionality. Therefore, the owner of technology could potentially license the use of his technology to as many licensees as he wishes, maximizing the earning potential of his technology constrained only by the terms of the agreements that he enters into with the potential licensees. In a sense, one technology could become the basis for a whole range of related or unrelated products and services made by one or many enterprises in a potentially large number of locations in one or many countries.

The traditional drivers of economic growth *viz.* land, labour and capital, are no longer sufficient to provide the necessary competitive advantage that makes the difference between companies that are otherwise very similar to one another. The answer lies in new or improved technology.

In selling or buying rights to the intellectual property in technology, the ownership rights for that technology pass from seller to buyer and it is a one-time activity. Here the legal transaction is called an "assignment". The technology is bought or sold for an agreed price. There will be only a few continuing obligations in the relationship between the seller (assignor) and the buyer (assignee). Generally, such transactions involve a one-time transfer of funds, but financial compensation might also be entirely or partially deferred and may depend on many factors or contingencies (such as the success of the commercialization). A technology owner normally has no experience in bringing a product to market and is not interested in being involved in such day-to-day matters as technology at work, may consider that an ideal solution would be to find a buyer for the technology and to complete the whole transaction at one time. In contrast, a licensing agreement transfers from the licensor to licensee the right to use the intellectual property in the technology and to make, use and sell products embodying the technology, in a specified manner for a specific time in a specified region. In other words, the licensor continues to have the proprietary rights over the technology and has only given a defined right to the use of that technology. In the field of biotechnology where transfer of technology alone may not be sufficient to practice the invention, the right to use (but not own) certain tangible property, usually biological material, may also be transferred through a patent license agreement. Licensing, therefore, entails very different legal and practical consequences to those of a sale or assignment. It also serves very different business purposes. If these purposes are not relevant for the parties then licensing is not the strategy to adopt.

Due diligence is a necessary step before embarking on any business transaction. It may include agreements on a multitude of other issues that generally linked to, but may be separate from the agreement to license technology. The technology may be protected by one or more patents, subject to copyright, trademark, design or trade secret. All of these issues may merit different agreements or perhaps constitute different parts of a single agreement. In these situations patent information on technological activity must be gathered from all the sources because technical solution to the problem may be found in a totally different technical field. If the technology is not protected, means it is in public domain then there is no issue of licensing of IPRs. If the technology has been protected then validity in the country and its maintenance must be looked for. It is worth mentioning that only some 5 million patents are in force out of 42 million patent documents. On an average for any one invention a patent application is filed in only four countries, which means there is a good possibility that a particular invention protected by a patent in one country may not be protected in many, most or all countries of interest to a prospective licensee. Further, it is possible that the effective use of a targeted patented technology depends on other patented technologies. This means that one or more licenses to use such other technologies would become necessary. Assessing all these issues will usually require the expert advice of an appropriately qualified intellectual property professional.

It is important to keep in mind that it is not sufficient to enter a negotiation based on pure trust as on many occasions the negotiations do not necessarily result in an agreement. To safeguard against such an eventuality, it is a standard practice to enter into a mutual non-disclosure agreement, also referred to as confidentiality agreement or a secrecy agreement. Any such agreement would have to be customized based on the facts and circumstances of a given situation and should be reviewed by an appropriate legal professional. If both parties have reason to believe that they are adequately prepared for the negotiation then the need for a preliminary understanding in the form of an MOU or Letter of Intent should normally not arise. However, despite the best efforts of the parties, there are situations in which it becomes necessary to enter into such an MOU or Letter of Intent prior to the signing of a licensing agreement. This may happen prior to the commencement of formal negotiations or sometimes during protracted negotiations when, for example, there is a need to publicly announce the launching of a new product or apply for funding. Before entering into an MOU or Letter of Intent it is important not to agree to anything proposed by the other side without understanding its implications for the final licensing agreement. This is particularly true in a country where an MOU or Letter of Intent is treated as legally binding.

In the agricultural research sector, public research institutions have the responsibility to see research through to commercialization in all but the few lucrative markets that attract the bulk of private-sector attention. Negative effects of IPR on non-profit 'commercialization' of innovations have until now been most apparent in the agricultural sector, and in non-profit institutions such as medical centres where clinical researchers wish to use patented diagnostic tools to treat patients, for a fee, as part of their research programme. It is the effect of IPR on the mission of these integrated enterprises, rather than on the environment perceived by the bench scientist, that is the key issue for the prospects for biotechnology innovations for agriculture in developing countries.

The number of biotechnology innovations which have been developed to the point of commercialization actions by public and non-profit agriculture is small. In the USA and some other developed countries, there is some evidence that university research projects designed to produce new crops with modern biotechnology have been shut down because of refusal of IPR-holders to permit commercialization of varieties incorporating their intellectual property. For example University of California researchers using a patented promoter of Life Science Corporation in the development of tomato variety genetically engineered to express endoglucanase gene to retard softening and improved shelf life characteristics. In another example, development of fungus-resistant strawberry at the University of California was blocked by lack of access to the necessary *Agrobacterium* transformation technologies. In the development of herbicide tolerant barley the owner of the relevant herbicide tolerance patent refused to negotiate commercialization rights, and indeed refused to discuss developing the germplasm itself. Likewise, there are similar reports of impediments to commercialization, in the form of refusal of freedom to operate, have been encountered in development of herbicide tolerant turf grass at the University of Michigan and of a herbicide tolerant lupin in Australia. The main point of these examples is not that they would all have been commercially successful given freedom to operate, but that freedom to operate was in these cases a serious barrier to a system of non-profit innovation that has responsibility for development to the point where they were made available to farmers in the field. Why do these happen? In economic terms, the 'transaction costs' must have been too high or perhaps the public-sector negotiators had unrealistic expectations regarding private sector largesse. It might be that the owner of key IPR is concerned with protecting itself from liability or from damage to its reputation due to misuse beyond its control. In some cases the expected financial gains, given the size of the market, might have been less than the cost in time and money to the IPR owner (public or private) of making and enforcing an agreement or perhaps the patent holder saw no reason to help out a potential competitor, for little financial return, in a market that could one day be of financial interest to the patentee.

There are evidences from surveys and case studies that there is a strong prima facie case for significant blocking effect of intellectual property claims in public/non-profit agricultural research that yields commercially attractive results. There have been cases in which US patents, later invalidated, have been used to hold up commercialization of products from developing countries. For example, yellow bean (enola bean) patented by a Colorado firm demanding licenses from importers of similar Mexican beans, Del Monte Fresh Produce warning against working on Pineapple plant material, though variety in question was not patented. These examples show that, even before TRIPS has had its full impact, confused perceptions of geographic scope of patents, its validity etc. may have a plausible discouraging effect.

When the rights to existing patents are needed to practice a technology, the dominant and overlapping patents claims must be examined because it can affect the right to use downstream innovations. For example Monsanto claim to the plant transformation method using *Agrobacterium* means that all patents in which the claims specifically depend on this transformation method are blocked by a previous patent US 6,369,298 a patent assigned to Pioneer Hi-Bred International.

1. Valorisation: Valuation of Technology

Unlike tangible property, which has well recognized means of establishing a value and thus a price, there is no easy way to determine the value of intangibles. However, as with any other transaction, a price must be established. Valuation of technology (valorisation) is a difficult exercise and often a subjective one. Valuing a technology becomes important when the potential licensee has recognized the need for new and most appropriate technology, identified the potential licensor and decided that a license arrangement is the most appropriate business strategy. Broadly, the worth of an IP/technology will be derived from the likely benefits that would accrue to its end-users, and the price will be determined from the extent of the benefits that the R & D agencies would deem to appropriate. Several methods can be used to value a technology. IP valuation may be subjective and depends on the data that is used in the valuation model, the valuations derived from each of the criteria will not be the same.

1.1 Cost Approach

The licensor's investment in the technology is represented by those costs associated with developing, protecting and commercialization of the technology. The goal should be for both the licensor and licensee to have a realistic understanding of the licensor's investment and its relevance to the payments to be made to the licensor by the licensee. The cost based approach value is based on historical or replacement cost. This gives however no indicator of future value

1.2 Income Approach

Anticipated future income that is attributable to the intellectual property (in the form of Trademark, Patent, Design, Copyright, etc.) is a key component of the income approach to value. Royalty rates are frequently used to drive the income approach to valuing IP. This calculates the amount IP owner would need to pay in royalty income if someone else owned the IP.

Successful technology licensing means, for the licensee, increased profits because of the use of IPR protected technology. Some licensing professionals start their valuation calculations with a rule of thumb, according to which the licensor should receive around one quarter to one third of the benefits accruing to the licensee, often referred to as the '25% rule'. By way of illustration, if a new product is expected to sell for INR 2 lacs, and all costs total INR one lac, there will be an operating profit of INR one lac. Of this, 25% is INR 25,000/-. This is the amount, according to the "rule", the licensor should receive, and could be a starting point for further negotiation having regard to the above risks and royalty variables and any other relevant factors.

It may be that one party does not wish to pay or receive running royalties for the term of the agreement, but wants only a lump sum (perhaps in time-based or event-based installments), and therefore a fully-paid-up license. In this event, the next step would be to prepare a statement identifying for each year all the cash inflows and outflows, for the term of the agreement (n), and to then apply the formula $1/(1 + r/100)^n$ and calculate the lump sum or Net Present Value (NPV). This calculation requires the selection of a discount rate, r, which is the cost of capital adjusted for risk and so effectively incorporates or reflects all the risks. The NPV establishes the present value of future income streams expected from the use of the technology under consideration. In these negotiations, one or both parties will hire accountants to run various scenarios of possible return and discount depending on certain scenarios. It should be noted that the NPV (also termed the Discounted Cash Flow or DCF) analysis is relevant to any issue where time and money are relevant factors. It can thus be a tool of wide application.

Income approach is based on future cash flow and economic profit. Future cash flow are however difficult to predict and therefore IP valuation remains risky.

1.3 Market Approach

It follows that comparable market transactions are a convenient and useful way of determining the value of asset in anticipation of negotiating a purchase or sale. An early survey by the Biotechnology Licensing Committee of the Licensing Executives Society (LES) reported that following ranges for non-exclusive licenses were considered representative for:

- Research reagents (e.g. expression vector, cell culture), 1-5% of net sales.
- Diagnostic products (e.g. monoclonal antibodies, DNA probes), 1-5% of net sales.
- Therapeutic products (e.g. monoclonal antibodies), 5-10% of net sales.
- Vaccines, 5-10% of net sales.
- Animal health products, 3-6% of net sales.
- Plant/agriculture products, 3-5% of net sales.

The following factors may be considered for agricultural technologies concerning SMEs and farmers with small and medium holdings in determining/ assessing the worth of IP/technology/know how and in fixing its price.

   i. Expected adoption level and expected benefits accruing to the end-users. Higher is the adoption rate and/or per unit benefit, higher will be the price e.g. the price of tomato seed may be higher than that of water melon or amaranth seeds. Similarly, price of a rice hybrid gaining popularity over a large area could be higher than a conventional rice variety.

  ii. Proportion of the benefits appropriated by the commercializing agency, where applicable: Higher are the benefits appropriated, higher will be the price, e.g., ready-mix baby food or other nutraceuticals.

 iii. Cost associated with up scaling/commercialization of IP: Higher is the cost of up scaling, lower will be the price, e.g., those plant based agro-chemicals or bio-agents that essentially need up-scaling to develop commercial product.

 iv. Impact on innovation market: Lower price may be charged for the IP/ technology which can increase competitiveness of the innovation market, e.g., indigenously adapted, modified laser leveler.

  v. End-users and impacts of IP: Low price may be charged for IP benefiting disadvantaged social groups (poor people, women, tribals, etc.) or increasing sustainability of natural resources, or protecting environment, e.g., varieties of underutilized crops and minor millets, and small tools for agricultural operations and harvesting/threshing like tubular maize sheller.

Institutions may take several different price norms in the market as basis for fixing the price of their IP. They may also consider fixing price cluster of technologies (e.g., hybrids, bio fertilizers, machinery, etc.) rather than fixing in individual cases; with a provision for different methodologies for different clusters. The institutions, instead of fixing a one-time price for the IP, can consider reviewing the price periodically; say once in three years, e.g., for breeder seed of vegetable and flower crops, nevertheless, if affirmative, this clause may be incorporated in the licensing contract/agreement.

Generalizations, surveys and industry norms at least provide a starting point. What can be much more useful, however, is knowledge of a comparable licensing arrangement in the same industry which could provide another basis or check for valuation of a particular technology. Usefulness of market approach is often limited and does not work because IP is unique and there are few comparable transactions which makes it difficult to set a price.

2. Licensing Agreement

License simply means permission granted by the owner of the intellectual property right to another to use it on agreed terms and conditions, for a defined territory and for an agreed period of time. Every license agreement is unique, reflecting the particular needs and expectations of the licensor and licensee. An infinite variety of agreements are possible, limited only by the needs of the parties and by the parameters of the relevant laws and regulations. However, certain issues are fundamental to the success of an agreement. i) License is the outcome of a business strategy and is a business relationship. Both the licensor and licensee must carefully consider whether entering into one or more licensing agreements fits into the business plan of the company, whether the expected revenues would be sufficient to justify the costs involved in engaging licensing activity and whether the financial terms make sense to both the parties. ii) A license agreement is a contract which means that legal requirements for a binding and enforceable contract are necessary. iii) The subject matter is intellectual property, which the licensor grants the licensee the right to use. Therefore, without intellectual property there is no technology licensing. iv) For effectively using the licensed technology a licensee has to access other technologies owned by another, which are proprietary. In these situations the licensee is obliged to obtain the rights to use the technology(ies) from the owner of the intellectual property right through a licensing agreement, which may be on a royalty free basis or negotiated on the basis of fair, reasonable and non-discriminatory terms.

Many license agreements involve a combination of one or more types of intellectual property. For example, patents and know-how license agreement, use of a trademark along with rights to make, use, sell, distribute and/or import a patented invention, a license may not mention a specific patent by number, but rather provide the specifications of a product and grant all IPRs necessary to manufacture and sell such a product. An agreement can include additional rights for carrying out further research or development or the provision of technical assistance.

In a license agreement the following are the important components:

2.1   Subject Matter

Subject matter is the first main section of the license agreement. It may include creations such as inventions, confidential information, the creativity expressed, business identifiers, etc. If license agreement involves computer software, then there may be specific clauses specifying the permitted use or application and requiring confidentiality to be maintained. Failure to clearly identify the subject matter of the license is a major pitfall. The parties should quote the patent number, clarify whether the license is to use software, documentation, a drug formula, a protocol, a text, a musical score, etc.

Prior to and during negotiations for a licensing agreement the licensor may have to disclose information which is considered confidential and should not be used or disclosed by the potential licensee. For example in the development of hybrid variety parental lines involved, male sterile lines, source of male sterility, etc. In the development of transgenic variety the concentration of growth regulators for regeneration of an explant, transformation protocol, use of specific promoter sequence or codon modification in the gene sequence for better expression of a gene. Hence, for the purpose of protecting the licensor's rights the following agreements can be signed prior to negotiations:

Confidentiality or Secrecy Agreement: Prior to and during negotiations for a licensing agreement the licensor may have to disclose information which is considered confidential and which should not be used by the potential licensee if the negotiation does not result in an agreement. For the purpose of protecting the licensor's rights, a confidentiality or secrecy agreement will often be signed by the parties as a condition precedent to disclosure and negotiation. The signing of a confidentiality agreement is also an assurance that the discussion is being entered into seriously.

Letter of Intent or Memoranda of Understanding: A Letter of Intent or a Memorandum of understanding (MOU) is a preliminary agreement that sets out the broad intentions of the parties in entering into a binding agreement. Such a Letter or MOU generally states that the parties have embarked on and intend to continue negotiations with the intention of concluding a license agreement. Preferably, it should indicate the period of time within which such an agreement is to be concluded. The legal consequences of such a Letter or MOU depend on the legal system in the country in question. Some national laws view them as legally binding, whereas others take the view that they establish the seriousness of intention of the parties but fall short of a binding contract. In any event, much will depend

Standstill and Related Agreements: In this Agreement, a potential licensor grants a potential licensee a period of time to consider entering into a licensing agreement with the licensor, and the licensor agrees not to entertain any other candidate until the expiry of that period. Such an agreement allows potential licensee flexibility in deciding whether to enter into a licensing agreement for the technology in question and, if so, some time to prepare for it. For example, researching the technological, financial, marketing and legal aspects of such a relationship. The licensor who provides a potential licensee with a Standstill Agreement is unable to grant other licenses for the period of the Standstill Agreement, which would normally mean a period of a few months.

Research Agreement: In research and development agreements, a research institution or company undertakes to carry out a research study or trials on the basis of its own existing expertise. The party providing the financial support for such a project is often a company seeking a technology focused outcome such as a new or improved process or product. A research agreement can be of particular interest to universities and companies in developing and least developed countries having expertise in areas that are specific to those countries, but that lack funding or other resources to undertake the necessary research and development. A partnership with a company that can provide the funding, complementary expertise and knowledge will create opportunities for knowledge sharing and for building up a research base vital in the modern knowledge economy.

Certain clauses in the agreement should take into account the following:

a. Define what is meant by confidential information.

b. Ensure that the licensee has or undertaken to put in place procedures for restricting the use of the information for the purposes as specified in the agreement and safeguarding it against disclosure.

c. Provide for liability in the case of accidental or negligent disclosure of the information to third parties who are not subject to the provisions of the license agreement and who are not otherwise informed of the confidentiality of such information.

d. Spell out the exceptions to the obligation, such as if the information is publicly available, that is, it is already known or has become known to the recipient in a legitimate manner or if it had been independently developed by the recipient.

e. Clarify as to how long these provisions will continue after the termination of the agreement and specify when the information should either be returned or destroyed.

2.2 *Extent of Rights*

It refers to the scope of the right, being exclusive, sole or non-exclusive, and the geographic territory for which the license is granted. The scope might also include improvements made to the technology during the license and will include the duration of the agreement.

The nature of the rights being licensed depends on the subject matter. For a patent, this would normally be the right to make, use and sell a patented product or use a patented process. In the case of a copyright license it may also include the right to reproduce, translate, display, modify and distribute. The rights might also be restricted according to a defined application or product. Thus, the licensed "field of use" for a vaccine might be the treatment of cancer, and there might be other licensees with rights for hepatitis and other diseases.

In a particular territory, the license may be exclusive, sole or non-exclusive.

A non-exclusive license, where the licensee is one of several licensees with whom the licensor has entered into agreements for the use and exploitation of the technology, is the preferred option of most licensors. By spreading the risks and rewards to several licensees, the licensor does not depend on the success of one licensee. He can maintain a better control over the technology and, by virtue of the fact that several licensees are using and exploiting the technology in several markets and perhaps in a variety of products, give the technology a chance to further evolve and develop.

An exclusive license usually describes the situation where the rights granted to the licensee even exclude the rights of the licensor in the territory. A sole license usually describes the situation where the licensor as well as the licensee can use the technology in the territory, but no one else can. This distinction can be blurred in practice and the term exclusive is sometimes used to mean what is really a sole license. In any event, under both types of license, the licensor is not permitted to grant other licenses (at least in the territory in which the license is expressed to be exclusive or sole). In that territory, the licensor is reliant on one licensee. Accordingly, it is important to ensure that the agreement contains appropriate incentives and/or penalties to protect the licensor in the event of non-performance by the licensee. These might include the payment of an annual minimum royalty. If the licensee does not make the required payment, then the penalties might be termination of the license or conversion of the exclusive license to a non-exclusive license.

If the license covers more than one territory, it may be exclusive in one while non-exclusive in another. The exclusivity may be limited, for example, to a field of use or period of time or linked to the achievement of milestones.

Where the license is non exclusive, the licensee may wish to include in the agreement a most favored licensee clause which in effect ensures that in the event that the licensor grants another licensee terms that are more favorable, then, by virtue of this clause, the present licensee would be entitled to terms as favorable as had been granted to the other licensee.

Territory: The extent of the license also refers to the geographic territory. For example, worldwide rights could be granted, or the rights could be for specific countries or even specific parts of countries (such as a state or region of a country).

Sub-license: The licensee, particularly if the licensee has an exclusive license, may wish to have the right to grant sub-licenses in its territory. If so, this needs to be specifically negotiated and stated in the agreement. It should also be stated if the licensor's prior written approval is required for the granting of any sub-licenses, the choice of sub-licensee and the conditions upon which such sub-licenses may be granted. Non-exclusive licensees are generally not granted the right to grant sub-licenses since a potential sub-licensee, could seek a license directly from the licensor.

Technical assistance: Depending on the kind of technology being transferred, there is often an agreement to provide the licensee with technical assistance in the form of documentation, data and expertise.

Term: The licensor may also wish to limit the term in order to assess the business efficacy of the licensee. The licensee may wish to extend the term if it is investing heavily in infrastructure necessary for exploitation of the intellectual property (e.g., a factory or a distribution channel). The only rule about the term of a license is that this depends entirely on the business needs of the parties and many tailored and negotiated outcomes are possible.

2.3 *Commercial and Financial*

An important factor in an agreement is commercial and financial considerations i.e. the valuation of the technology. Payments to the licensor for the acquisition and use of technology are usually classified as lump sums and royalties, and many agreements contain both types of payment.

Lump sums are payable on the happening of a particular event. There may be one sum only, payable on signing the agreement. If there were no further payments, this would be considered a fully-paid-up license. Time-based payments are certain in that the amounts are known and agreed, and they are risk-free which will be paid when the specified period has elapsed. No further action is required by the licensee or the licensor. Performance-based payments, on the other hand, depend on the occurring of certain events, such as the first commercial sale. As the payments are not made if the event in question does not occur, it is important to clearly define events such as first commercial sale, etc.

Royalties are regular payments to the licensor, which reflect the use of the technology by the licensee. As they link use with a monetary amount they can be a good reflection of the value of the technology to the licensee and, accordingly, royalties are the most usual type of payment in license agreements. Royalties have two key components: the royalty base and the royalty rate.

The royalty base could be the cost of manufacturing or the profit from selling the licensed products or calculated from the licensees sales. This could be the number of units of the licensed product sold with the licensee paying a fixed amount of, say, INR 100 per unit. All that needs to be ascertained is the number of units sold, and the royalty payable is determinable. Alternatively, the royalty base could be either the gross or the net sales receipts of the licensee.

Second key component of royalties is the royalty rate. Negotiation of the royalty rate is fundamental to the success of the agreement. Too high a rate can mean the license is unprofitable for the licensee while too low a rate mean the licensor does not receive an adequate return, which might lead to reduced expenditure on continuing research and development. Either of these adversely affect the relationship between the parties and the success of the agreement.

The other factor is royalty variable. One approach is that the royalty rate reduces as the volume increases or time passes. Thus, a royalty rate of 15% might reduce to 12.5% after the sale of one million units, then to 10% after five million units. This might be on an annual or a cumulative basis. The reverse is also possible, with the royalty rate increasing as the volume increases. The first approach has the objective of encouraging the licensee to increase production and hence the royalties payable to the licensor. The reverse approach imposes lower royalty costs on the licensee at the beginning while the technology is being introduced and sales are low and increases them as market share is gained. Another possible approach is that the licensee is required to pay the licensor an annual minimum royalty. Thus, the sum of Indian Rupees 10 lacs might be payable for year 2 of the license, increasing to Indian Rupees 15 lacs for year 3 and Indian Rupees 20 lacs for each year thereafter. This is particularly appropriate where the license is exclusive and the licensor needs to ensure that minimum royalties are received. If they are not, the licensor is free to work with another partner so that his technology and intellectual property rights are not wasted by poor exploitation. The reverse is also possible, and instead of there being a continuing annual minimum royalty, the license can become "paid up" or royalty free. This would happen when an agreed event occurred, such as, for example, fifteen years of commercial production and/ or total royalties paid reaching an agreed total sum, whichever event occurs first. This has the objective, after the licensor has been substantially rewarded, of ensuring that the licensee is rewarded as well.

The issue of inflation is effectively provided for where the royalty rate is expressed as a percentage of sales. Where, however, the royalty is a specific amount in a specified currency, it is usually reviewed regularly, say, annually or every two years, and adjusted, if the national law so permits.

The financial administration provisions of the license agreement includes obligations on the licensee to keep accounts and records, to report the results and pay the consequent royalties. The royalty reports, which might be required once, twice, or four times a year, might need to be certified by the licensee's chief financial officer or auditor. In any event, the licensor usually reserves the right to inspect, or have a third party inspect, the licensee's accounts and records. Financial administration also includes where the parties are from different countries, the issues of currency and taxation. The currency of payment is not always the currency in which royalties arise. In these cases, it will be necessary to specify when the conversion is to be made and the rate to be used.

2.4 Infringement

When all or part of the technology has the benefit of patent or other intellectual property protection, it is important to provide for what will happen if there is any infringement. There are two situations where infringement could occur.

The first is where a third party is using the protected technology but does not have a license. Here the licensee is facing competition and is likely to be at a financial disadvantage as the infringing competitor is not paying royalties. The licensee, particularly if he is a non-exclusive licensee, will expect the licensor to take steps to deal with the infringement. For instance, the licensor could negotiate with the third party so that it becomes a licensee. If this is not appropriate or is not successful, then the licensor may need to take legal action. Until proceedings have been instituted, the license agreement might provide that the licensee has the right to pay royalties into a separate bank (escrow) account, which are paid to the licensor when proceedings are instituted. If, however, proceedings are not instituted within, say, three years, then the accrued royalties could be returned to the licensee and, thereafter, the license could be royalty-free.

The second infringement situation is where a third party claims that the licensee is using technology in respect of which the third party has obtained protection. In this situation, the licensee may be faced with the prospect of not being able to continue to use all or some part of the licensed technology. Again the licensee will look to the licensor to provide support and assistance. However, the licensor might argue that it is the licensee who has control over the application of the technology and that, in any event, before signing the agreement and commencing production, the licensee should have carried out the relevant searches, which would usually have revealed the presence of these pre-existing rights. Even so, the license agreement might provide that the parties would ascertain whether it is possible for the licensor to provide non-infringing technology. If not, the issue is whether the third party's patent is valid, and, if so, the licensee might require the licensor to obtain a license from the third party and a consequent adjustment to the financial arrangements between the licensor and the licensee.

2.5 Product Liability

Product liability can have important financial consequences. The risk is that there might be injury or damage, to person or property, arising from a licensed product that is defective. The need is to identify the source of a potential defect and to assign responsibility accordingly. Thus, the licensee would usually be responsible for any manufacturing defects or for inadequate quality control. The licensor may supply components to the licensee, and, in this event, the licensor would usually be responsible for any defects in those components.

3. Licensing Rights in Agricultural Technologies

For agricultural technologies non-exclusive license is the preferred one. Licensor can maintain a better control over the technology and also by virtue of the fact that several licensees are using and exploiting the technology and It does not affect the livelihood and also remains competitive. The idea is that these can lead to wider adoption of technologies; maximizing research benefits to farmers and other end users. There may be flexibility in fixing the license fee. For example, it may be low (e.g. Four lacs INR in first instance, which may increase to 5-6 lacs INR in the second instance in case of higher demand or vice versa). There is less likelihood that a single firm will have adequate capacity and marketing infrastructure to cover the entire country, including the remote and far-flung areas. Therefore, non-exclusive licenses by government agencies with respect to agricultural technologies on regional/ area basis will enhance the local availability of the technology and reduce the transportation cost and thereby market-price. When the license is non-exclusive, the licensee may wish to include most favoured licensee clause in the agreement.

Exclusive license can be issued in exceptional cases like (i) commercialization in foreign countries, (ii) difficult areas offering low incentives; (iii) commercialization requiring high development cost, iv) exclusive license should cover only one territory while it is non-exclusive in another, etc. The exclusivity may be limited to a field of use or period of time or linked to achievement of milestones. The duration, whether limited or indefinite, for which such licenses are issued, will depend upon market conditions. A specific sub-licensing clause shall be negotiated and incorporated particularly in the exclusive licenses, which may require the other contracting parties to share a part of the license fee and/or royalty from the sub-licenses that they may enter into. In case a client insists on the exclusive license, (i) it should be negotiated at a high license fee and/or royalty offer, and (ii) negotiations should be made for offering such license for a limited period (3 or 5 years) after the expiry of which re-negotiations should re-occur to account for current demand/scope of IP.

Sometimes in an exclusive license, the licensee wish to have the right to grant sub-licenses in its territory, then it needs to be specifically negotiated and stated in the agreement and prior approval of licensor is required. Non-exclusive licensees are generally not granted the right to grant sub-licenses since a potential sub-licensee could seek a license directly from the licensor.

Agreement for joint commercialization of IP can be entered into by institutions in cases where (i) a close scientific supervision is required, (ii) commercialization is done using the institute resources; (iii) technology is extended under scientist entrepreneurship; or (iv) any other such situation.

Asian countries economies are dependent upon agriculture and governments must consider the following with respect to commercialization of its plant varieties.

i. Advance breeding material or parental lines shall not be transferred/ licensed on exclusive basis but these should be first registered with Bureaus of Plant Genetic Resources in their respective countries or at Asian regional bureaus before any transfer/licensing deal is to be negotiated/ entered into.

ii. Breeder seed: To maintain the quality of seed supplied to farmers either one time transfer or recurrent supply of breeder seed of every licensed variety will be a 'must' and the licensor institutions should ensure as per the terms of the licensing contract/agreement with the licensees.

iii. The license fee and/or sale price of breeder seed and royalty either on a fixed basis or through negotiations with the licensee, as appropriate, may be fixed for each variety considering the cost of seeking and maintaining the plant variety right, cost of production, handling and supply of breeder seed and other institutional costs on equitable basis.

4. Management of IPRs and Technology Transfer

Management of IPRs requires capacity building in the countries as per their needs. Capacity building is never monolithic in nature but a multidimensional and complex activity. Capacity building should be in all the areas viz. IPR management, information and documentation, patent search and analysis, techno-legal drafting of patent applications, patent litigation, licensing, valuation and negotiating IP licensing deals. No exercise at a national level can succeed if all or most players from innovators to entrepreneurs, scientists to students, NGOs to farmers are not engaged in the activity. IPRs are often considered synonym of patents or at best patents, trademark and copyrights. Sometimes people even use word 'patent' as a substitute for 'protect'. There is a need to adopt different means for awareness such as contact programmes, workshops, trainings, print media, bulletins, internet, videos, etc. Awareness by itself is of little use if the State does not create and provide suitable systems to enable scientists, technologists, industrialists, farmers to protect their rights. These means would be in terms of technical guidance, financial support, legal help and other facilitation steps. Capacity building has to be multifaceted at the national level, regional level and at multi-country level which are in the same stages of development so as to remain ahead or at par in the knowledge race.

Government departments like Atomic Energy, Space, DRDO and R&D agencies like CSIR, ICAR have their in house system for looking after their needs of IPRs from capacity building to filing to commercialization. CSIR, ICAR as well as other central R&D organizations have developed a well structured Technology Management Units which facilitate filing of IPs in India and abroad and commercialization of technologies.

Various funding agencies for R&D must spell out in their policy decision who will be the owner of IPs generated by their funding and how to move about for commercialization. The national R&D organizations, IITs, universities, industries whether public or private must have their IPR policies.

The Development of skills and competence to manage IPRs and to leverage its influence should be given a major thrust. This area calls for significant technological insights and legal expertise and should be handled differently from the present, and with high priority. Efforts should be made for synergism between industry and scientific research by creating Autonomous Technology Transfer Organization as associate organization of universities and national laboratories to facilitate the transfer of know-how generated to industry.

The Department of Science and Technology (DST) set up the Patent Facilitating Centre (PFC) at the Technology Information Forecasting and Assessment Council (TIFAC) in 1995 as a small initiative to address the need of awareness creation among scientists, helping them to protect their inventive and original work through IP laws. PFC is one such system available in the country which provides full technical, legal and financial support though their Patent Information Centres (PICs) in different States of India for protection of inventions emanating from educational institutions, schools, colleges and government departments.

4.1 Innovations Related Incentives

An innovative industry can gain competitive advantage in the market if it develops the necessary expertise and skills in developing and manufacturing new products, which are patented. Saha (2005) has suggested that the following incentives would be extremely useful in promoting the culture of innovation and intellectual protection in industries and academic and R&D institutions.

1. Excise duty waiver on patented products for a certain period of time from the date of commencement of commercial production.

2. Financial government support for commercialization of indigenous technologies.

3. Exemption from drug price control order – Bulk drugs produced based on indigenous R&D may be exempted from drug price control for a certain period of time from the date of commencement of commercial production.

4. Weighted tax deduction on R&D expenditure – R&D expenditure should be available to companies engaged in the business of biotechnology, agricultural technologies, manufacture of agrochemicals, etc. The expenditure on scientific research shall include expenditure incurred on clinical trials, field trials, obtaining approvals from the regulatory authority of state/province and central governments and for filing a patent application.

5. Accelerated depreciation allowance – Depreciation allowance at a higher rate should be made available to the industries, which are involved in the manufacture of goods or products based on indigenous technologies.

6. Tax holiday to R&D companies for some years which are involved in the development of agri-technologies.

7. Income tax relief on R&D expenditure.

8. Tax deduction for sponsoring research.

Selected reading

Chawla, H.S. 2007. Managing Intellectual Property Rights for Better Transfer and Commercialization of Agricultural Technologies. J Intellectual Property Rights, 12: 330-340.

Dubey Rounak and Dubey Rupal. 2014. Valorisation of Intellectual Property in Knowledge base economy. In: Contemporary Management Strategies in Intellectual Property Rights-Relevant to NAM and other Developing Countries (Eds. Sarah Norkor Anku, Olufolake Sola Davies and Rungano Karimanzira), Astral International Pvt Ltd., Delhi, pp. 300-309.

Exchanging Value – Negotiating Technology Licenses, A Training Manual by World Intellectual Property Organization and International Trade Centre, WIPO Publication No. 906(E), 2005.

ICAR Guidelines for Intellectual Property Management and Technology Transfer/Commercialization, ICAR, New Delhi, 2006.

Saha, R. 2006. Management of intellectual property rights in India, paper presented at Workshop on IP Management in Public Private Partnership, Manesar and Bangalore, 2006, 9-35.

PART 2

IPR Protection and Conservation
of Biological Resources

CHAPTER 11

Plant Variety Protection

Plant variety protection relates to intellectual property rights over plant varieties, which guarantee the rights holders' exclusive commercial rights for a specific period of time. The issue of plant variety protection (PVP) or plant breeders rights (PBR) was brought into world-wide focus by the agreement on TRIPs which is a part of establishing of WTO in 1995. The process of plant variety protection recognizes new creations in plant breeding by vesting monopoly rights on the plant breeders who created them.

The development of private sector in agriculture field in 20th century led to increasing demand for a form of intellectual property rights protection over plant varieties to give sufficient incentive to the private sector to enter the seed business.

Plant breeders first sought protection under the industrial patent system. However, a number of technical difficulties were encountered in applying the rules of patent system to plant varieties. First, plant material was not regarded as capable of meeting the requirements of novelty, inventive step and usefulness. Second, it was not thought to be in the public interest to permit such an extensive monopoly over plant varieties, given their societal importance. Underlying view was that it was desirable to retain as far as possible, the tradition of free exchange of new plant material between plant breeding institutes. This would ensure the widest new combinations of genetic information.

1. Historical Perspective

The Paris Convention of 1883 is the first multilateral agreement for harmonizing intellectual property laws. It extended protection to industrial property, which was recognized to also apply to agricultural and extractive industries and to all manufactured or natural products. First attempt to recognize the intellectual property right of plant breeder was the enactment of the Plant Patent Act by USA in 1930. This Act allowed for patenting of asexually reproduced cultivars (except tubers). These rights were extended to new and distinct asexually varieties for a period of seventeen years.

By 1960s some European countries enacted the Plant Breeders' Rights Laws under the UPOV Act which ultimately led to UPOV 1978 and 1991 Acts and has been discussed below. At the same time several attempts were made to enact similar protection in the United States, including a proposal to revise the Plant Patent Act to include sexually reproduced plants. These early attempts were unsuccessful. The Plant Variety Protection (PVP) Act was enacted on December 24, 1970 by USA which provided protection for sexual reproduction in plants, including seed germination. With this Act most commercial crops were now protected by patent laws for seventeen years but it was limited by two major exemptions: seed saving by farmers and for research purposes. Under the PVPA "brown bag" exemption, farmers could continue to save, replant and resell protected seeds to other farmers. Thus, two divergent views emerged; one was to extend patent protection to plants while the other was to extend the *sui-generis* protection, recognizing plant breeder's right. The question of *sui generis* intellectual property right protection for plant varieties has become a matter of great importance following the adoption of TRIPs agreement. *Sui generis* is a Latin word meaning 'its own kind/genus or unique in its characteristics or being the only example of its kind or constituting a class of its own'. All it means that it allow a country to develop a *sui generis* legal system to protect plant varieties in a manner that it suits to its socio-economic political ambience and is effective in the protection such offered.

A General Agreement on Tariffs and Trade (GATT) was established in 1947 (came into force on 1 January 1948) to deal with post-world war scenario concerning multilateral trade issues, particularly with a view to serve as means of stimulating world trade in goods, remove/minimize impediments to trade among member countries, and to create a level playing field. In Uruguay the 8th round of such GATT negotiations started in 1986 but it was concluded on April 15, 1994 at Marrakesh, Morocco with the signing of an accord by 123 countries that led to the formation of World Trade Organization (WTO) on 1st January 1995. This round was significant because agriculture was included as a tradable commodity and agreement on TRIPs was achieved.

The extension of intellectual property protection to agriculture raised few concerns. First concern is to protect the interests or the privilege of farmers to save repeatedly and use for sowing their farm saved seed of protected varieties as an exception to the plant breeder's right, and secondly protect the right of farmers in terms of getting equitable share of benefits derived from the use of plant genetic resources conserved and preserved by them. The first approach is reflected in the UPOV Act, the European Nations' effort towards securing the plant breeders' right, while the second approach is reflected in the FAO International Undertaking on Plant Genetic Resources, which was renegotiated and adopted as the International Treaty on Plant Genetic Resources for Food and Agriculture (ITPGRFA) known as Seed Treaty.

2. UPOV

On December 2, 1961, Five European countries (Belgium, France, the Federal Republic of Germany (FRG), Italy and the Netherlands) agreed in a Diplomatic Conference to provide *sui-generis* IPR protection to plant varieties and formed the Union for the Protection of New Varieties of Plants, original in French 'Union International Pour la Protection des Obtentions Vegetales' (UPOV). UPOV came into force only in August 1968 after the UK, the FRG and the Netherlands had ratified it. Of the five original signatories, France ratified it and joined the Union in 1971, Belgium in 1976 and Italy in 1977. The UPOV underwent three amendments in 1972, 1978 and in 1991. The 1978 Act came into force in 1981 and the 1991 Act in April 1998. There are at present two main UPOV Acts of 1978 and 1991 under which the parties to the convention have joined. Countries are not obliged to join UPOV as a result of their affiliation with any other organization or the ratification of any specific treaty. Membership is purely voluntary. The Convention requires member countries to provide an intellectual property right specifically for plant varieties. The UPOV Convention provides a *sui generis* form of intellectual property protection which has been specifically adapted for the process of plant breeding and has been developed with the aim of encouraging breeders to develop new varieties of plants. UPOV provided for the rights of plant breeders, and prohibited from extending two or more types of protection to a particular plant species.

ASSINSEL spearheaded non-patent type IPR on new plant varieties. While the USA took to the patent route, the Europe evolved the *sui generis* system to grant IPR on new plant varieties. Two main reasons were suggested for the inappropriateness of patent system to protect plant varieties: i) Plant material was not regarded as capable of meeting the requirements of novelty, inventive step and disclosure in patent specification unlike in the case of inventions in physical sciences; ii) It was viewed that permitting such an extensive monopoly possible under patent over plant varieties is undesirable in public interest, for the important reason to retain within the IPR system the tradition of free exchange of new plant material among plant breeders and their institutes to ensure the widest possible dissemination and use of new combination of genetic information.

The 1978 Act entered into force in 1981 while the 1991 Act came into force on April 24, 1998. Both the 1978 and the 1991 Acts set out a minimum scope of protection and offer member states the possibility of taking national circumstances into account in their legislation. Some countries have ratified the 1991 Act whereas others the 1978 Act. The convention has 70 countries. The 1991 Act entered into force on 24 April, 1998 and on that same date the 1978 Act was closed to future accessions except by a few states already in the process of adhering to it. PBR registration systems are broadly the same in each of the UPOV member countries. Some differences may arise in the scope of protection and the administrative procedures in applying for and progressing an application. Not every UPOV member protects every species. UPOV Acts of 1978 and 1991 have been summarized in Table 11.1.

India and so many other countries do not protect plants by strict patenting system. But there is a mandate in the TRIPs agreement that plant

Table 11.1: A comparison between the two UPOV Acts on plant variety protection

Sr. No	Issue	1978 Act	1991 Act
1.	Membership	Only a state can be the party	Intergovernmental organization competent for enacting and implementing legislation with binding upon on all its member states can also be the party
2.	Discovery	Breeder is entitled to protection as discoverer irrespective of the origin, artificial or natural of the variation	A mere discovery is not sufficient. The breeder must also have developed the variety
3.	National treatment	Member state may limit the right on a new variety to national of states which also apply that Act. A similar reciprocity rule may also be applied by a member state granting more extensive rights	Reciprocity rule does not apply. Operation of the principle of national treatment to one and all without qualification

4.	Scope	Authorization of breeder is required for
		  i. the production for purposes of commercial marketing of the propagating material;
		  ii. the offering for sale of the propagating material;
		  iii. the marketing of such material;
		  iv. the repeated use of the new plant variety for the commercial production of another variety (e.g. hybrids);
		  v. the commercial use of the ornamental plants or parts thereof as propagating material in the production of ornamental plants or cut flowers;
		  vi. does not require authorization for use of the material for further research;
		  vii. farmers can use/reuse his produce as seed and can dispose off his farm produce.
			Authorization of breeder is essential for
			  i. production or reproduction;
			  ii. conditions for the purpose of propagation;
			  iii. offering for sale;
			  iv. selling or other marketing
			  v. exporting;
			  vi. importing;
			  vii. stocking.
			Furthermore, the 1991 Act specified four subject matters to which the breeder's right extends
			  i. the protected variety itself
			  ii. varieties which are not clearly distinguishable from the protected variety;
			  iii. varieties which are essentially derived from the protected variety;
			  iv. varieties whose production requires repeated use of protected variety.

5.	Minimum number or species to be covered	At least 5 to start with At least 10 within 3 years At least 18 within 6 years At least 24 within 8 years	UPOV 1978 member states, all after 5 years transitional period. If only bound by 1991 Act to start with 15 plant genera/species. All genera and species after 10 years.
6.	Period of protection	18 years for grapevines and tress including rootstocks 15 years for all others	25 years for grapevines and tress including rootstocks 20 years for all others

Contd...

Table 11.1: (Contd.)

Sr. No	Issue	1978 Act	1991 Act
7.	Prohibition on dual protection with patent	Yes, for same botanical genus or species.	No.
8.	Breeders' exemption	Mandatory. Breeders free to use protected variety to develop a new variety.	Permissive, but breeding and exploitation of new variety "essentially derived" from earlier variety require right holder's authorization.
9.	Farmers' privilege	Implicitly allowed under the definition of minimum exclusive rights.	Allowed at the option of the member state within reasonable limits and subject to safeguarding the legitimate interests of the right holder.

varieties must be protected. In pursuance to the TRIPs agreement, India has enacted "Protection of Plant Varieties and Farmers' Rights" (PPV&FR) Act, 2001, a *sui generis* system of plant variety protection. This law is unique which has brought forth the farmers rights under the gambit of law. The model for this was the UPOV Act. India is not a member of UPOV. Under the UPOV a plant variety qualifies for protection when it meets three essential criteria, (i) distinctness, (ii) uniformity and (iii) stability, and the variety should be new in commercial sense. Application for its protection can be filed in the country where developed or in any other UPOV member country.

The 1978 amendment of UPOV left the provision of farmers' privilege unaltered. However, UPOV 1991 Act, made the farmers' privilege optional to the member countries. This provision allows farmers to use the product of the harvest of the protected variety, which they may obtain by planting on their own holdings, for further propagating purposes. The Convention requires that the farmers' privilege be regulated "within the reasonable limits and subject to the safeguarding of the legitimate interests of the breeder".

3.   Functions of UPOV

The convention has two main functions:

1. Prescribe minimum rights that must be granted to plant breeders by the member countries, in other words, it specifies a minimum scope of protection.
2. Establish standard criteria for grant of protection.

4.   The UPOV 1978 and 1991 Acts

The purpose of UPOV convention is to ensure that the member states of union acknowledge the achievement of breeders of new plant varieties by making available to them an exclusive property right on the basis of a set of clearly defined principles. It provides the standard conditions for the grant of protection and excludes the impositions of any other additional conditions. If a legal right is to be granted in respect of the unit plant material that constitutes a variety and if that right is subsequently to be effectively enforced, the identity of the variety must be established. The 1978 UPOV Act adopts most of the international IPR obligations including a definition of the applicable subject matter and protected material, eligibility requirements, exclusive rights, national treatment, reciprocity, terms of protection and exceptions and limitations to exclusive rights. It does not, however, contain any provisions on most favoured nation (MFN) treatment or enforcement. The UPOV Convention also establishes a multilateral system of national treatment, under which citizens of any member state are treated as citizens of all member states for the purpose of obtaining plant breeders rights. Application for its protection can be filed in the country where developed or in any other UPOV member country. There are key differences in UPOV 1991 Act from UPOV 1978 in three areas: (a) the coverage of varieties qualifying for protection, (b) the nature of rights enjoyed by the breeders and, (c) the rights over "essentially derived varieties" (Table 11.1). UPOV 1978 permits its signatories to protect plant varieties either with a distinct breeder's right or with a patent. However, this condition does not exist in 1991 Act.

The UPOV Convention is not self-executing. The UPOV Act does not become enforceable in domestic law until the state enacts a national plant variety protection law that conforms to the Act's requirements. Thus, each member state must adopt legislation consistent with the requirements of the convention and submit that legislation to the UPOV Secretariat for review and approval by the UPOV Council.

4.1 Conditions for the Grant of Protection

The four eligibility requirements for a specific variety – novelty, distinctness, uniformity and stability – are same in both the 1978 and 1991 Acts. Breeders' right on a variety qualifies for protection when it meets the following essential criteria:

i) **New or Novelty (N)** –Variety is said to be new in the commercial sense that a variety must not have been commercialized prior to certain date established by reference to the date of application for protection. A variety must not have been sold, or otherwise disposed of, in territory of the member of the Union concerned (contracting party) for more than one year prior to the application for as breeders right or more than four years (six years for trees or of vines) in the territory of another member of the Union.

ii) **Distinctness (D)** – Variety is considered to be distinct from existing, commonly known varieties by one or more identifiable morphological, physiological or other characteristics (Art. 6(1)(a)). The Guidelines for the Conduct of Tests for Distinctness, Homogeneity (uniformity) and

Stability (UPOV Guidelines) use both qualitative and quantitative plant characteristics, including such visible attributes as leaf shape, stem length and colour, to determine if the difference between varieties is "clear and consistent."

iii) **Uniformity (U) or Homogeneous** – Under the 1978 UPOV Act, a variety has to be "sufficiently homogeneous, having regard to the particular features of its sexual reproduction or vegetative propagation" (Art. 6(1)(c)). The UPOV Guidelines further clarify that to be considered homogeneous, the variation shown by a variety must be "as limited as necessary to permit accurate description and assessment of distinctness and to ensure stability."

iv) **Stability (S)** – The stability requirement is a temporal one, requiring the breeder to show that the essential characteristics of its variety are homogeneous or uniform over time, even after repeated reproduction or propagation (Art. 6(1)(d)). In practice, what has been shown to be homogeneous is usually considered to be stable as well. In other words its relevant characteristics remain unchanged after repeated propagation or, in the case of a particular cycle of propagation at the end of each such cycle. The variety must be stable in appearance and its clonal characteristics over successive generations under the specified environment.

A unique and unambiguous denomination (name of the new variety) in accordance with the provisions is to be given for protection of a variety.

4.2 Varieties Covered by the Breeders' Right

Under the UPOV Convention the breeders' rights extends to:

i. The protected variety

ii. Varieties not clearly distinguishable from the protected variety

iii. Varieties whose production requires repeated use of the protected variety

v. Essentially derived varieties

5. Plant Breeders' Right

Plant breeders' right (PBR), also known as plant variety rights (PVR), are rights granted to the breeder of a new variety of plant that give the breeder exclusive control over the propagating material (including seed, cuttings, divisions, tissue culture) and harvested material (cut flowers, fruit, foliage) of a new variety for a number of years.

The UPOV 1978 Act requires its signatories to protect a variety's reproductive or vegetative propagating material but does not require protection of harvested material or other marketed products, with the exception of ornamental plants that are used for commercial propagating purposes (Art 5(1)). Like its predecessor, the 1991 Act recognizes the right of breeders to use protected varieties to create new varieties. However in the 1991 Act extensive additions to the 1978 Act were made. The breeder's prior authorization must be obtained for the use of reproductive or vegetative propagating material of the variety for (1) production or reproduction, (2) conditioning for the purpose of propagation, (3) offering for sale, (4) selling or marketing, (5) exporting, (6) importing and (7) stocking for any of these purposes mentioned in (1) to (6). In addition the UPOV contains an optional provision to extend the scope of the breeders' right to products made directly from harvested material, where this has been obtained through the unauthorized use of protected variety, unless the breeder has had reasonable opportunity to exercise his right in relation to the harvested material. The above provisions shall also apply in relation to varieties which are essentially derived from the protected variety, where the protected variety is not itself an essentially derived variety.

6.   Duration of Breeders' Right

Under UPOV 1978 Act the protection is granted to new varieties of all genera and species of plants, for a period 18 years for trees and vines and 15 years for all others. The protection granted for the new variety authorizes the breeder with the exclusive right to commercially exploit the variety by direct sale or by licensing to others for sale.

The 1991 Act extends the term of protection to 25 years for trees and vines and 20 years for all other varieties.

7.   Exceptions and Limitations to Breeders' Rights

Following are the major exceptions and limitations to exclusive rights exist under the 1978 Act: (1) a breeders' exemption and (2) a farmers' privilege.

7.1   Breeders' Exemption

Under PBR regime, the use of material of protected variety (*the initial variety*) for the development of new varieties is exempted from protection. The PBR for these new varieties will be of the breeder who developed them, and the holder of PBR title of the initial variety will have no claim to it. This provision is called *breeders' exemption*. Under the UPOV 1978 Act, all new varieties evolved using protected varieties were exempted from protection under this provision.

UPOV 1991 Act has somewhat limited the scope of breeder's exemption by bringing 'essentially derived varieties' under the cover of PBR protection granted to the initial variety. An essential derived variety is defined as a variety predominantly derived from another initial variety, which retains the expression of essential characteristics from the genotype or combinations of genotypes of the initial variety except for one or few distinguishable characteristics. As a result of this modification, a breeder who inserts a single gene (e.g. disease resistant gene) in to a protected variety will now have to obtain the permission from the original right holder before marketing the new variety.

However, the breeder's right shall not extend to (i) acts done privately and for non-commercial purposes, and (ii) acts done for experimental purposes.

7.2   Farmers' Privilege

Under the UPOV 1978 Act PBR system generally allows the farmers to use the material of a protected variety harvested on their farm for planting of their new crop without any obligation to the PBR title holder. This exemption is usually referred to as *farmers' privilege*. However, the scope of this so-called farmers' privilege varies widely. Some nations only permit farmers to plant seeds saved from prior purchases to be used on their own land holdings, while others allow them not only to replant but also to sell limited quantities of seeds for reproductive purposes, a practice often referred to as "brown bagging."

But in the UPOV 1991 Act, this farmers' privilege is limited in scope and has been made 'optional' and each UPOV member state can either allow or disallow this privilege. UPOV 1991 Act deprives the farmers of its rights to use, reuse their produce as a seed. Although farmers are broadly exempted from the breeders monopoly for non-commercial use of their produce from a protected variety including propagating another crop from harvested material on their own farm. It should be clearly understood that unlike 1978 Act, the 1991 version of farmers' privilege does not authorize farmers to sell or exchange seeds with other farmers of protected varieties produced on their farms for propagating purposes. This limitation has been criticized as inconsistent with the practices of farmers in many developing nations, where seeds are exchanged for purposes of crop and variety rotation.

8.   Patent Protection v/s PBR

A number of governments in the industrialized world, including the United States, Japan, Australia, New Zealand, Sweden and the United Kingdom, have capitalized on this opportunity by permitting plant breeders to obtain patent protection in new varieties provided that the eligibility requirements for a patent have been met. While UPOV Acts developed on the basis of *sui generis* system of protection gave plant variety protection rights comparable to a patent. Plant breeders rights are related to the rights available to a patentee, where the model of protection is plant based.

There are significant differences in approach between plant breeders' rights and patents. In the case of plant breeders' rights, the eligibility requirements for protection are not stringent and accordingly the scope of protection granted is quite narrow, both in terms of exclusive rights and the various exceptions and limitations to those rights. While, eligibility requirements for patent are high and difficult to meet, but once granted a patent conveys broad rights to exclude third parties from exploiting the patented invention. Depending on the needs and level of development of plant breeder industries within its territory, a government may decide that either or both forms of protection will provide the appropriate incentives to encourage plant-related research and innovation.

To avoid confusions between patent protection and plant variety protection by PBR, a comparison of the two UPOV Acts with patent is presented in Table 11.2.

9.   Plant Variety Protection – USA

In the USA there are three main ways in which an inventor or breeder may obtain formal IPR on plant material:

1. Plant patent under the Plant Patent Act (PPA) 1930

2. Utility patent, under the Utility Patent Act (UPA)

3. Plant breeder's rights through the Plant Variety Protection Act (PVPA), 1970

A plant patent is granted by the US Government to an inventor who has invented or discovered and asexually reproduced a distinct and new variety of plant, other than a tuber propagated plant or a plant found in an uncultivated state. The grant, which lasts for 20 years from the date of filing the application, protects the inventor's right to exclude others from asexually reproducing. This protection is limited to a plant in its ordinary meaning:

Utility patent may be granted in the U.S. for any new plant in which man has had "a hand" in the creation thereof. The first utility patent on plant was given in 1985 to Tryptophan overproducer mutants of cereal crops (US Patent No. 4,642,411) referred as Hibberd case. Following the principle established in the Chakrabarty case, it was decided that normal US utility patents could be granted for other types of plants also e.g. genetically modified plants.

Plant breeder's rights through the Plant Variety Protection Act (PVPA), 1970 provided protection for sexual reproduction in plants including seed germination. Its purpose is to "encourage the development of novel varieties of sexually reproduced plants" by providing their owners with exclusive marketing rights of them in United States. Fungi, bacteria and first generation hybrids are excluded from the PVP protection. Varieties sold or used in the United States for longer than one year or more than 4 years in a foreign country are also ineligible for protection. A Certificate of Protection remains in effect for 18 years from the date of issuance. The Act was limited by two major exemptions: seed saving by farmers and for research purposes. Under the PVPA "brown bag" exemption, farmers could continue to save, replant and resell protected seeds to other farmers. The PVP Office is responsible for administering the PVP Act. It is organized within the Agricultural Marketing Service of the U.S. Department of Agriculture.

TABLE 11.2: Comparison of principal differences between plant variety protection under UPOV 1978 Act, UPOV 1991 Act and TRIPs-compatible patent laws

Subject	Breeders' rights in UPOV 1978 Act	Breeders' rights in UPOV 1991 Act	TRIPs-compatible patent laws
Object	Plant variety	Plant variety	Invention
Documentary examination	Required	Required	Required
Field examination	Required	Required	Not required
Eligibility conditions for protection	Plant varieties that are novel, distinctive, uniform and stable.	Plant varieties that are novel, distinctive, uniform and stable.	Plant varieties, plants, seeds and enabling technologies that are novel, involve an inventive step and industrial application.
Minimum exclusive rights in propagating material	Production for purposes of commercial marketing; offering for sale; marketing; repeated use for the commercial production of another variety.	Production or reproduction; conditioning for the purposes of propagation; offering for sale; selling or other marketing; exporting; importing or stocking for any of these purposes.	Making the patented product, using the patented process or using, offering for sale, selling or importing for those purposes the patented product or the product obtained by the patented process.
Minimum exclusive rights in harvested material	No such obligation, except for ornamental plants used for commercial propagating purposes.	Same acts as above if harvested material obtained through unauthorized use of propagating material and if breeder had no reasonable opportunity to exercise his or her right in relation to the propagating material.	Making the patented product, using the patented process or using, offering for sale, selling or importing for those purposes the patented product or the product obtained by the patented process.
Breeders' exemption: use of protected variety for breeding further varieties	Mandatory. Breeders free to use protected variety to develop a new variety.	Permissive. No authorization required except for "essentially derived varieties".	Generally not recognized. Require authorization of the patentee of the right holder
Farmers' privilege: Use of propagating material of the protected variety grown by a farmer for subsequent planting on the same farm	Implicitly allowed under the definition of minimum exclusive rights.	Permissive within reasonable limits and subject to safeguarding the legitimate interests of the right holder.	Require authority of the patentee.
Additional exceptions to exclusive rights	None specified.	Acts done privately and for noncommercial purposes, acts done for experimental purposes.	Research and experimentation. All exemptions must comply with three-part test of TRIPs article 30.
Minimum term of protection	18 years for trees and grapevines; 15 years for all other plants.	25 years for trees and grapevines; 20 years for all other plants.	20 years from date the patent application filed.

10. Plant Variety Protection in India: Protection of Plant Varieties and Farmers' Rights Act, 2001

India is signatory to WTO agreements and it has to abide by the TRIPS regulations. As per article 27.3(b) of the TRIPs which demand that member countries should protect their plant varieties either by patent, or an effective system of *sui generis* protection, or a combination of these two. In this context India chose a *sui generis* system for protection of plant varieties. An Act named as Protection of Plant Varieties and Farmers' Rights (PPV&FR) Act 2001 was enacted in India on October 30, 2001. The rules under the Act were notified on September 12, 2003. Central Government established the PPV&FR Authority on 11th November, 2005 with its Head Office located at Delhi. The PPV&FR Act is TRIPs compliant and compatible with UPOV system of plant variety protection. But India is not a member of UPOV Convention.

11. Plant Variety

A plant variety is an assemblage of cultivated individuals which are distinguished by any character (morphological, physiological, chemical or any other) significant for the purpose of agriculture, forestry or horticulture and which when reproduced (sexually or asexually), or reconstituted, retain their distinguishing features.

As per PPV&FR Act, 2001 a 'variety' means a plant grouping except microorganism within a single botanical taxon of the lowest known rank, which can be:

(i) defined by the expression of the characteristics resulting from a given genotype of that plant grouping;

(ii) distinguished from any other plant grouping by expression of at least one of the said characteristics; and

(iii) considered as a unit with regard to its suitability for being propagated, which remains unchanged after such propagation, and includes propagating material of such variety, extant variety, transgenic variety, farmers' variety and essentially derived variety.

To promote the development of new varieties of plants and to protect the rights of farmers and breeders, The PPV&FR Act 2001 provides protection to following types of plant varieties:

12. Types of Plant Varieties Protected

i. Newly bred varieties.

ii. Extant varieties – The varieties which have been notified under Indian Seeds Act, 1966 and have not completed 15 years as on the date of application for their protection.

iii. Extant-Farmers' varieties– The varieties which have been traditionally cultivated, including landraces and their wild relatives which are in common knowledge, as well as those evolved by farmers. [For example: Indrasan, Hansraj and Tilakchandan (Rice); Rampur local (Sorghum); Kudrat 9 (Wheat)].

iv. Extant- Variety of Common Knowledge – A variety which is not released and notified under the Seeds Act, 1966 but is well documented through publications and is capable of satisfying the definition of variety, or has become a matter of common knowledge and the variety is under cultivation or marketing during the time of filing of application for registration. [For example: KMH-25K55, KMH-25K60, SYN-CO-6661, Vivek Sankul Makka-11(Maize); JK Vijay (Wheat), MECH12Bt (Cotton), Ankur Rupali (Rice)].

v. Essentially derived varieties- In respect of a variety (the initial variety), shall be said to be essentially derived from such initial variety when it is predominantly derived from such initial variety, while retaining the expression of the essential characteristics that results from the genotype or combination of genotypes of such initial variety; is clearly distinguishable from such initial variety; and conforms (except for the differences which result from the act of derivation) to such initial variety in the expression of the essential characteristics that result from the genotype or combination of genotype of such initial variety. [For example: VICH-5 BGII (Tetraploid cotton)].

vi. Transgenic varieties.

To facilitate the registration of plant varieties, Authority has opened two branch offices of the Plant Varieties Registry also, one at Birsa Agricultural University, Ranchi and other at Assam Agricultural University, Guwahati. These branch offices will function within its territorial limits and will also keep a copy of National Register of Plant Varieties.

13. Registration of Plant Varieties

All the varieties will be registered with PPV&FR Authority. DUS guidelines for different plant species which includes cereals, pulses, oilseeds, flowers, spices, vegetables, fruit trees, ornamentals and medicinal and aromatic plants have been notified by PPV&FR Authority in the Gazettes. The registration of varieties is now open for 102 plant species. Once notified, application may be filed for registration of varieties under the category of new plant varieties, essentially derived varieties (EDV), extant varieties (notified under the Seeds Act 1966), extant (varieties of common knowledge and farmers' variety). The Act has laid down the norms for registration of plant varieties, fee structure, provisions of opposition, DUS testing of material, etc. If any farmer or association of farmers is applying for registration of a plant variety then this category is not required to pay any fee for either registration or DUS testing. An affidavit for Rs 100/- on non judicial stamp paper has to be submitted with the application form indicating that the variety does not contain any GURT or terminator gene technology. A flow diagram for registration of plant varieties in shown in Figure 11.1.

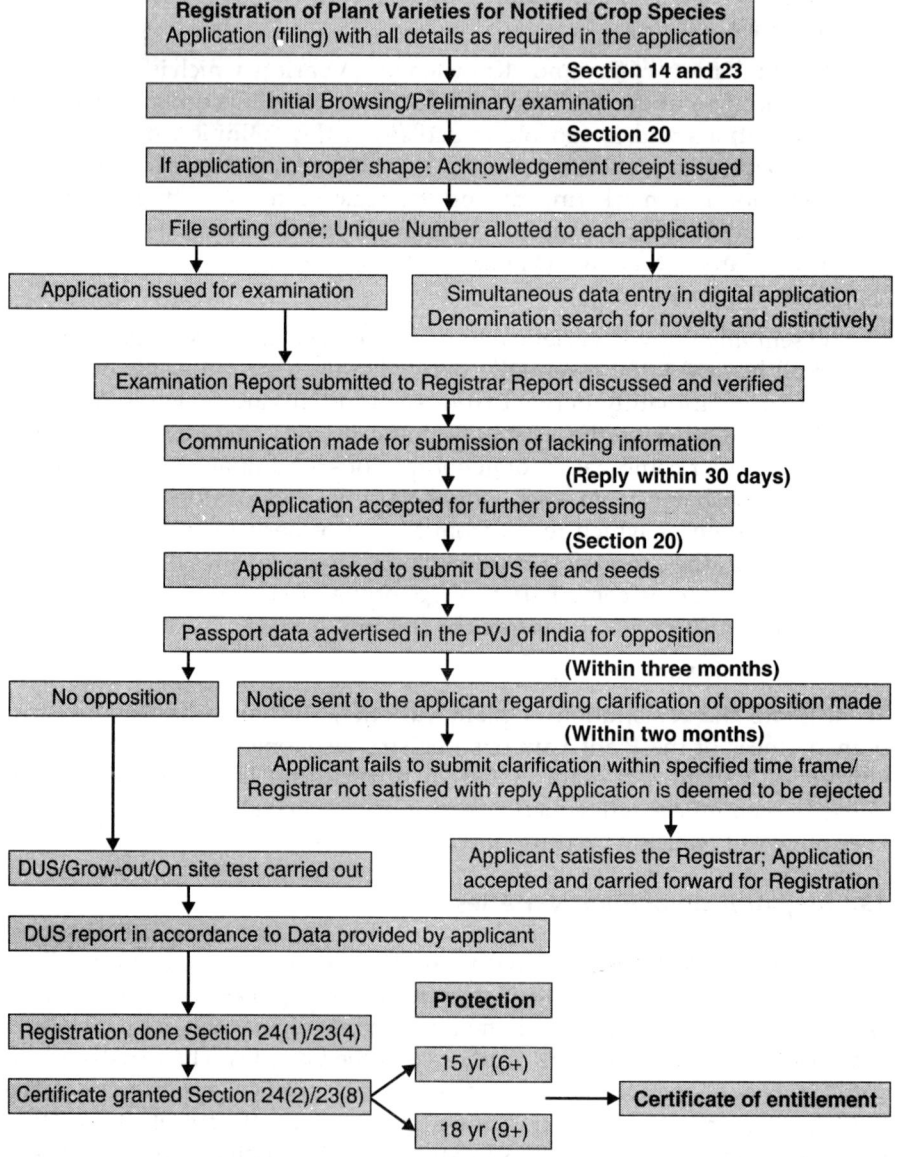

Fig. 11.1: Flow diagram of plant variety registration procedure

13.1 Eligibility Criteria for a Variety to be Registered

For a new variety to be eligible for registration, it must conform to the criteria of novelty, distinctiveness, uniformity and stability (NDUS [Section 15 (1)–

(3)]. However, an extant variety shall be registered under this Act within a specified period if it conforms to criteria of distinctiveness, uniformity and stability (DUS) as shall be specified under the regulations.

Novelty (N): It is the basic requirement for registration of newly developed plant variety. A variety is novel, if, at the date of filing of application for registration for protection, the propagating or harvested material of such variety has not been sold or otherwise disposed of in India, earlier than 1 year and outside India (in case of trees and vines earlier than six years, or in other cases, earlier than four years).

Distinctiveness (D): any plant variety which is clearly distinguishable by at least one essential characteristic from any other variety whose existence is a matter of common knowledge in any country at the time of filing of the application.

Uniformity (U): newly developed plant has to give uniform characters subject to the variation that may be expected from the particular features of its propagation but it is sufficiently uniform in its essential characteristics. Uniformity standards vary in different crops depending upon their nature of mode of reproduction as self or cross pollinated.

Stability (S): essential characteristics of newly developed variety remains unchanged after repeated propagation or, in the case of a particular cycle of propagation, at the end of each such cycle. Basically it means the characters are genetically inherited and are not due to environmental influences.

13.2 Varieties that Cannot be Protected Under the PPV and FR Act

Varieties that shall not be granted protection under the PPV & FR Act are (Section 29):

1. The varieties whose commercial exploitation may affect protection of public order or public morality; human, animal or plant life and health or may cause serious prejudice to the environment.
2. The varieties whose genus/species is not notified in the Official Gazette at the time of filing application.
3. The varieties which involve any technology that is injurious to life and health of human beings, animals and plants and which includes Genetic Use Restriction Technology (GURT)/terminator technology.

13.3 Things to be Kept in Mind for Making an Application for Registration of Variety

For registration of a plant variety the following points should be kept in mind:

1. Denomination assigned to such variety.

2. Accompanied by an affidavit that variety does not contain any gene or gene sequence involving terminator technology.

3. Complete passport data of parental lines, geographical location in India and all such information relating to the contribution, if any, of any farmer, village, community, institution or organization in breeding, evolving or developing the variety.

4. Characteristics of variety with description on novelty, distinctiveness, uniformity and stability.

5. A declaration that the genetic material for breeding has been lawfully acquired.

6. A breeder or other person making application for registration shall disclose the use of genetic material conserved by any tribal or rural families in the breeding or development of such variety.

The application for registration of a variety is to be made in the form as prescribed by the PPV & FR Authority. There is an application form (Form I) for registration of new variety, extant variety and farmer's variety and another form (Form II) for essentially derived variety and transgenics. A technical questionnaire form is also to be filled up giving all the details of the concerned variety. These filled application forms must be accompanied by the fee prescribed by the Authority.

13.4 Applicant for the Registration of a Plant Variety:
Any of the Following can Apply for Registration of a Variety

1. any person claiming to be the breeder of the variety;

2. any successor of the breeder of the variety;

3. any person being the assignee of the breeder of the variety in respect of the right to make such application;

4. any farmer or group of farmers or community of farmers claiming to be breeder of the variety;

5. any person authorized to make application on behalf of farmers and

6. any university or publicly funded agricultural institution claiming to be breeder of the variety.

14. DUS Testing

DUS is the criteria on the basis of which the Breeders' Rights will be granted to a variety by the Authority. DUS test will be used as one of the main criteria for deciding the novelty of a variety. Present system of DUS testing involves the comparison of candidate variety with the existing referred varieties by recording the phenotypic characters, which are (mostly) morphological and physiological in nature. Usually the DUS examination required more than one independent growing cycle with reference to ecosystem of the variety for studying the consistency of results. DUS testing will be conducted for two years for self-pollinated crops but cross-pollinated crops may require three years (Table 11.3). For some crops such as fruit trees, the same plants are examined over successive years. In some circumstances authorities can allow only one growing season. The distinctness of self-pollinated crops can be established using characters which can be assessed by visual examination and whose expression falls into clearly defined discrete states. In cross-pollinated crops many of the varietal characteristics are on a continuous scale of expression and require measurement. Usually, distinctions can be determined only on the basis of statistical analysis in cross pollinated crops. Different DUS testing centres have been identified for crops and notified in the Gazette.

For different crop species tolerable limit varies for uniformity of characters. For example the off-types limit of 4/1500 in rice, 2/1000 in wheat, 3/100 in Maize inbreds and S.C. hybrids are permitted. For farmers varieties off-types limit is doubled.

Table 11.3: DUS testing of varieties

Type of variety	Type of test	No. of locations	No. of seasons
New	DUS test	2	2
Variety of common knowledge	DUS test	2	1
Farmers'	Grow out test	2	1
Essentially derived variety	Manner of testing shall be decided by Authority on case to case basis		

To qualify for registration under the act, a new variety has to conform to the criteria of novelty (NDUS). While for extant varieties which includes varieties notified under Seeds Act, 1966, farmers' varieties and varieties of common knowledge (VCK), DUS criteria is to be followed. The application of registration of an EDV shall be accompanied by all the relevant documents alongwith other details. The Authority has constituted a six member expert committee which will act as an advisory body to the Authority for evaluation and recommendation of application filed. Once the Registrar is satisfied regarding the requirements, then it will be referred to the expert committee which will suggest the tests and procedures for establishing that the variety is derived from an initial variety. The rights of a breeder of a variety or an EDV are same provided that the authorization by the breeder of the initial variety to the breeder of EDV may be subject to such terms and conditions as both the parties may mutually agree upon.

The Authority has started the registration of varieties under this category also. The DUS testing shall be field and multi-location based for at least two crop seasons and special tests will be laboratory based. There shall be an option on the matter of DUS testing that a panel of three experts shall visit the On-Farm test sites for two similar crop seasons.

When DUS testing fails to establish the requirements of distinctiveness, then special tests mechanism has been provided in the Act. Special test has to be laboratory based and the Authority shall charge separate fees for the special tests which are to be identified on certain set principles. Broadly, these tests can be classified into five main groups: physical, biochemical, molecular, organo-leptic and response tests. To begin with Authority has constituted a task force for identifying special tests for cotton, rice, oilseed, wheat, maize and medicinal and aromatic plants.

15. Rights

15.1 Breeders Rights

The certificate of registration for a variety issued under this Act shall confer an exclusive right on the breeder or his successor or his agent or licensee, to produce, sell, market, distribute, import or export of the variety [Section 28 (1)]. Breeder shall enjoy provisional protection of his variety against any abusive act committed by any third party during the period between filing of application for registration and decision by the Authority. Breeders' rights would not apply in case when farmers save, exchange or use a part of the seed from the first crop of plants which they have grown for sowing on their own farms to produce a second and subsequent crops. Plant breeders would also not be able to exercise their rights in case where plants or propagating material of the protected varieties is used as initial source of variation for the purpose of developing new plant varieties.

15.2 Researchers' Rights

The researchers have been provided access to protected varieties for conducting experiments or research and use of a variety as an initial source of a variety for the purpose of creating other varieties. In case a registered variety is required as a parental line for commercial production of newly developed variety then authorization from the breeder of the registered variety is required [Section 30].

15.3 Farmers' Rights

Indian law follows a holistic approach. The *sui-generis* system adopted by India is unique in the world in the sense that it has taken farmers' rights concept a step forward and genuinely addresses the concerns of farmers as breeders, innovators, conservers, etc. It has tried to incorporate the features of UPOV and International Treaty on Plant Genetic Resources for Food and Agriculture (ITPGRFA) also known as Seed Treaty along with certain distinctive features of its own as per requirement of farmers. It is pertinent to note that the Act recognizes the farmer as a cultivator, conserver and breeder. This embraces all farmers, landed or landless, male and female. Under the Sec. 2(k) of PPV&FR

Act, a farmer means any person who; i) cultivates crops by cultivating the land himself; or ii) cultivates crops by directly supervising the cultivation of land through any other person; or iii) conserves and preserves, severally or jointly, with any person any wild species or traditional varieties, or adds value to such wild species or traditional varieties through selection and identification of their useful properties. PPV&FR Act of India recognizes various rights of farmers as per Section 39 (Table 11.4).

Table 11.4: Rights Provided to Farmers in PPV&FR Act, 2001 of India v/s ITPGRFA and UPOV

Type of rights	Description of rights	Available in ITPGRFA or UPOV
Farmers' rights	1. Rights to seeds	ITPGRFA, UPOV
	2. Right to register varieties	UPOV
	3. Right to reward and recognition as conserver	ITPGRFA
	4. Right to Information about expected per- formance and compensation for under-performance	–
Other rights	5. Right to Benefit Sharing	ITPGRFA
	6. Right to compensation for undisclosed use of traditional varieties	ITPGRFA
	7. Right to adequate availability of registered material	UPOV
	8. Right to free services	–
	9. Protection from innocent infringement of breeders' rights.	–

Indian PPV&FR Act allows farmers to save, use, sow, resow, exchange, share or sell his farm produce including seed of a variety protected under this Act, but it prohibits that the farmer shall not be entitled to sell branded seed of a variety protected under the Act [Sec. 39, 1(iv)]. PPV&FR Act distinguishes from UPOV Act which treats this as farmers' privilege rather than right. Exemption for farm saved seeds by farmers to save, use and exchange seed but not sell seed without penalty under plant breeders right system are referred as farmers' privilege. Since the use and exchange of saved seeds was considered non-commercial and hence was considered outside the scope of Plant Breeders' Rights. In the PPVFR Act the farmers have been given the right to register farmers varieties themselves [Sec. 39,1(i)]. The Act treats the farmer as plant breeder so far as farmers' variety is concerned and they can register them under the Act without paying any fess. Farmers have the right to claim compensation for under performance of a protected variety from the promised level [Sec. 39(2)], benefit sharing for use of biodiversity conserved by farming community [Sec. 41]. According to the concept of benefit sharing, whenever a variety submitted for protection is bred with the possible use of a landrace, extant variety or farmers' variety, a claim can be referred either on behalf of the local community or institution for a share of the royalty [Sec. 41(1)]. Farmers' have also been excluded from paying any fee in any proceeding before the Authority or Registrar or Tribunal or the High Court.

15.4 Community Rights

It is compensation to villagers or local communities for their significant contribution in the evolution of a variety which has been registered under the Act. Any village or local community in India can claim the credit for the contribution to a particular plant variety registered as a new plant variety. Any person or group of persons or any governmental or non-governmental organization may on behalf of the people of the village or community in India, can file in any notified centre, claim their contribution in the evolution of a variety [Section 41]. After verification, if the Authority is satisfied, and after giving an opportunity to the breeder to file an objection and of being heard, subjected to the limit notified by the Central Government, it may by order grant such compensation to be paid to the claimant. Authority can direct breeder of a variety to deposit compensation (arrear of land revenue) to the Gene Fund.

16. Compulsory License

In the Act a provision of compulsory license has also been put. According to this, after the expiry of three years from the date of issue of certificate of registration of a variety, any person interested can claim in an application to the authority alleging that reasonable requirements of the public for seeds or other propagating material have not been satisfied or that the seed or other propagating material is not available to the public at a reasonable price and pray for the grant of a compulsory license to undertake production, distribution and sale of the seed or other propagating material of that variety [Sec. 47(1)].

17. Certificate of Registration

Once the variety has been tested for its features then the Registrar of the Authority will issue the certificate of registration. It shall have the validity of nine years initially in case of trees and vines with renewal up to a period of 18 years. For other crops certificate of registration will be issued for six years initially with renewal up to 15 years. In case of extant varieties the validity period is 15 years from the date of notification of that variety by the Central Government under section 5 of the Seeds Act 1966 [Section 24(6)]. The Authority has also opened a "National Register of Plant Varieties" having all details of the registered plant varieties and kept at the Headquarters of the Authority at New Delhi. This Register is an authentication of the plant breeders rights granted to the applicants. As a requirement under the Act, for the purpose of benefit sharing, the Authority shall also send a copy of the certificate of registration to the National Biodiversity Authority and Indian Council of Agricultural Research.

18. Surrender, Opposition and Revocation

Surrender: A breeder of a variety registered under the Act, may at any time by giving notice in the prescribed manner to the Registrar, can offer to surrender his certificate of registration.

Opposition: Any person within three months from the date of advertisement of an application for registration may file an opposition based on the grounds provided in the Act. Both opponent and applicant file their pleading and evidence and the opposition is finally heard by the Registrar. If the opposition is allowed then the applicants cannot proceed further with the registration. In case the opposition is rejected then the variety proceeds for registration.

Revocation: Revocation for registration is decided by the Authority on application filed by any interested person on the grounds provided in the Act. No revocation is done without offering an opportunity of hearing to the registered breeder. If revocation is allowed by the Authority then the certificate of registration becomes invalid. Any person aggrieved with decision of the Authority or Registrar in an opposition or revocation proceeding may file an appeal to the Tribunal and subsequently to the Higher Court.

19. Infringement and Penalty

If a person infringes the rights of the registered breeder in respect of registered variety or registered denomination without the permission of breeder then it constitutes infringement.

Conditions lead to infringement of the rights: Any of the following may be a case of infringement under the PPV&FR Act:

1. If a person who is not a breeder of a variety registered under the Act or a registered agent or a registered licensee of that variety, sells, exports, imports or produces such variety without the permission of its breeder or within the scope of a registered license or registered agency without their permission of the registered license or registered agent.

2. If a person uses, sells, exports, imports or produces any other variety giving such variety, the denomination identical with or deceptively similar to the denomination of a variety registered under this Act in such a way that it causes confusion in the mind of general people in identifying the registered variety.

Any of the following acts applied to a registered variety can be considered as false denomination:

1. If applied to the variety itself.

2. If applied to the package in which the variety is sold or possesses such package.

3. If he uses the registered variety for any purpose of trade or production.

4. If the denomination describes or designates the use of variety or its propagating material similar to the registered variety.

5. If used for any advertisement, invoice, catalogue, business letter, price list or other commercial document for a variety and such variety is delivered to a person who made a request or order.

Penalty: Any person who applies any false denomination to a variety or indicates the false name of a country or place or false name and address of the breeder of a variety registered under the Act shall be punished with an imprisonment which is initially for three months and may be extended to two years, or with a fine of rupees fifty thousands which may extend to five lakhs or both imprisonment and fine can be imposed to the offender.

Any person sells, or has in possession for sale or for any purpose of trade or production of any variety having false denomination or applies false indication of the country or place in which such variety was made or produced or the name and address of the breeder of that variety, in such case he will be punished with an imprisonment of not less than six months which is extendible up to two years or with a fine not less than fifty thousands rupees which may extend to five lakhs, or imprisonment and fine.

If any person makes representation with respect to the denomination of a variety or its propagating material or essentially derived variety or its propagating material as a registered variety then he shall be punished with imprisonment for a term not less than six months which may extend to three years or with a fine not less than one lakh which may extend to five lakhs, or with both.

20. National Gene Bank and Field Gene Bank

As per the Act it is mandatory to maintain the seed samples/ propagating material of registered plant varieties up to a period of protection provided to the candidate variety and also to address the issues for intellectual property of plant varieties including legal requirements such as infringement of plant breeder's rights, compulsory license, etc. Authority has established the National Bank at Old Campus of National Bureau of Plant Genetic Resources (NBPGR), New Delhi for medium term storage of true samples of orthodox seed of all registered varieties for their entire period of protection. The seed samples kept in the National Gene Bank at low temperature (3-5°C) so as to maintain genetic purity, viability and health during the period of protection beyond which the denomination and variety may go under public domain. After the expiry of protection period, seed material may be submitted to NBPGR/ any public repository.

For perennial plants (fruit trees and plantation crops) such as mango, citrus, eucalyptus, popular, rubber, coffee, etc. which either produce 'recalcitrant (which are either short lived or do not withstand desiccation) seeds to no seeds at all, clonally propagated and have long regeneration cycles or sexually sterile, 'Field Gene Bank' is a practice worldwide as an effective *'ex-situ'* conservation strategy. Such field gene bank are developed in places concerned, where suitable agro-climatic conditions like soil, water, area being relatively free from disease/pest infestation are available. For collection and maintenance of varieties released (referral collection) of perennial crop species collected from different niches so as to preserve sub species /intra-varietal variability at one place, Authority has established four Field Gene Banks at Dapoli, Maharashtra (for tropical and sub-tropical crops), Ranchi, Jharkhand (Eastern ecosystem), Mashobra, H.P. (for temperate crops) and Jodhpur, Rajasthan (for arid ecosystem).

21. National Gene Fund

The Central Government has constituted a Fund called the National Gene Fund which would be enriched through the benefit sharing received in the prescribed manner from the breeder of a variety or an essentially derived variety registered under the Act, or propagating material of such variety or essentially derived variety: the annual fee payable to the Authority by way of royalty by the breeders of the registered variety; the compensation deposited in the Gene Fund under sub-section (4) of section 41; the contribution from any national and international organization and other sources. The Gene Fund shall, in the prescribed manner, be applied for meeting any amount to be paid by way of benefit sharing under sub-section (5) of section 26; the compensation payable under sub-section (3) of section 41; the expenditure for supporting the conservation and sustainable use of genetic resources including *in-situ* and *ex-situ* collections and for strengthening the capability of the Panchayat in carrying out such conservation and sustainable use and the expenditure of the scheme relating to benefit sharing framed under section 46 of the Act.

22. Benefit Sharing

The Act provides for benefit sharing involving registered varieties. It first applies specifically to EDVs. In the second, any village local community can claim benefit for contributing to the development of a variety registered under the Act [Sec 41]. For a variety registered as EDV, any person or group of persons, being citizen(s) of India or firm or governmental or non-governmental organization formed or established in India, within a period of six months from the date of publication of the contents of the certificate of registration, can claim a share of benefits that may arise from its commercialization on behalf of any village or local community. The Authority shall establish the justification of the claims and determine the amount to be paid as benefit share on the basis of two criteria: a) the extent and nature of the use of genetic material of the claimant in the development of the variety for which benefit sharing has been claimed; and b) the commercial utility and demand in the market for the variety. The amount of benefit sharing, if any, would have to be deposited in the National Gene Fund by the breeder of the variety. In the second circumstance, any person or group of persons, being citizen(s) of India or firm or governmental or non-governmental organization formed or established in India can make a claim on behalf of a village or local community for the contribution that they had made in the evolution of any variety registered under the Act. If, upon investigation, the claim was found justified, after the breeder was given an opportunity to file objection and to be heard, the Authority shall, by order, determine the amount of benefit sharing to a variety taking in to account the criteria: a) the contribution of the claimant in selecting, conserving and providing the genetic material, b) the contribution of such genetic material in providing one or more traits which conferred high commercial value to the variety, and c) the contribution of such genetic material to impart high combining ability to the parents of the hybrid variety relating to benefit sharing. Amount of compensation as the Authority deems fit would be deposited by the breeder in the National Gene Fund.

23. Plant Varieties Protection Appellate Tribunal

The Act provides for establishment of Plant Varieties Protection Appellate Tribunal (PVPAT). All orders or decisions of the Registrar or Authority relating to registration of variety/ registration as an agent or licensee can be appealed in the tribunal. Further all orders or decisions of Authority relating to benefit sharing, revocation of compulsory license and payment of compensation can be appealed to the Tribunal. The Tribunal consists of one judicial member and one technical member. The form of appeal and period within which it must be preferred has been prescribed in PVPAT (Application and Appeals), Rules 2010. There is a transitory provision by which it is provided that till PVPAT is established, Intellectual Property Appellate Board established under section 83 of the Trade Marks Act 1999 shall exercise the jurisdiction of PVPAT. The decisions of PVPAT can be challenged in High Court. The Tribunal shall dispose of the appeal within one year.

Selected Reading

Chawla, H.S. 2009. Introduction to Plant Biotechnology (3rd edition). Science Publishers Inc., USA and Oxford & IBH Publishers, Delhi, India, pp. 728.
Chawla, H.S. 2014. Protection of Plant Varieties, Farmers' Rights and Benefit Sharing. In: Contemporary Management Strategies in Intellectual Property Rights-Relevant to NAM and other Developing Countries (Eds. Sarah Norkor Anku, Olufolake Sola Davies and Rungano Karimanzira), Astral International Pvt Ltd., Delhi, pp. 270-290.
The Protection of Plant Varieties and Farmers' Rights Act, 2001 along with The Protection of Plant Varieties and Farmers' Rights Rules 2005, Universal Law Publishing Co., Delhi, 2015.
http://www.plantauthority.gov.in
http://www.planttreaty.org
http:// www.upov.int/en/publications/tg_rom/introduction.html
http://www.upov.int/en/publications/conventions/1991/act1991.
Singh Anshu Pratap, Padmavati Manchikanti and Chawla, H.S. 2011. *Sui generis* IPR laws *vis-a vis* Farmers' rights in Asia and their implications under WTO. *J Intellectual Property Rights* 16 (2): 107-116.

CHAPTER 12

Biodiversity

The TRIPs agreement includes a number of forms of IPRs with implications on biodiversity conservation and for plant variety protection. The interest of gene rich developing countries is best served by global action aimed at conservation, protection of plant genetic resources by IP laws, sustainable use of biodiversity, sharing of benefits from the use of bioresources and to prevent biopiracy of genetic resources. Biopiracy means the practice of commercially exploiting naturally occurring biochemical or genetic material, especially by obtaining patents that restrict its future use, while failing to pay fair compensation to the community from which it originates. However, it is necessary to analyse what the genetic resources rich countries are doing to protect their plant genetic resources and giving benefits to the farming community (Table 12.1).

Table 12.1: A comparison of developing and developed countries on biodiversity related aspects

Developing Countries	Developed Countries
Biodiversity rich	Relatively poor biodiversity
Vavilovian Centres of Diversity	Nil to almost nil
Indigenous knowledge	Largely non-existent
Biodiversity supported by cultural diversity	Largely non-existent
Genetics, breeding and biotechnology base poor	Rich base in technology
Largely *in situ* conservation	Largely *ex situ*, but *in situ* for their own non agricultural biodiversity
Conservation indigenous-science-based	Largely modern science based
Largely subsistence or intensive agriculture	Largely modern science based
Sustainable utilization of biodiversity	Capacity exists
Poverty	Rich base
Largely bio industrial development	Largely industrial development

Biodiversity refers to the variety of life forms-that different plants, animals, and microorganisms, the genes they contain and the ecosystem they form-that make up the fabric of the earth's biosphere. Biodiversity is recognized to be maintenance of ecosystems and organism generally as well as providing essential services for human's survival.

Environment means all those components that surround us, influence us and affect us in some way or other. Environment essentially constitutes all the surrounding physical and biological factors with which a given organism interacts.

Since all these are inseparably connected with human beings, they are also linked with the way we live and interact with them.

1. Introduction

Biodiversity includes the variability amongst living organisms from all sources including, inter alia, terrestrial, marine and other aquatic ecosystems and the ecological complexes of which they are part; this includes diversity between species and of ecosystem. In essence, biodiversity represents all life.

Indian subcontinent is among one of the 12 mega biodiversity centres and represents two of the eight Vavilovian centres of origin representing diversity of crop plants. The Indian Gene Centre (Vavilonian "Hindustani Centre") is considered the centre of origin and domestication of as many as 356 major and minor crop plant species and as many as 326 wild relatives of crop plants. With only 2.4 percent of the land area, India accounts for 7-8% of the species recorded so far in the world. India has about 126,756 species of plants, animals, fungi and microorganisms already identified and classified. It is estimated that the floristic spectrum of India comprises of over 30,000 species (excluding fungi) of which the flowering plants with about 17,500 species constitute the dominant group representing about 7% of the flowering plant species of the world. Endemism in Indian flora is now almost well documented. [Endemic species means a species is restricted to a particular geographic region as a result of factors such as isolation or in response to abiotic conditions]. It is estimated that out of 17,500 species of flowering plants, 140 genera and 5285 species are endemic to the country. The Zoological Survey of India has recorded over 81,000 species of animals. The Himalayan mountains system covers only 18% of the geographical area of India, but accounts for more than 50% of India's forest and 40% of the species endemic to the Indian subcontinent. It is suggested that roughly 1.75 million described species of organisms may be only around 10% of the total life forms.

2. Loss of Biodiversity

It is generally accepted that the current loss of ecosystems, species and gene pools is faster than at any time since the extinction of the dinosaurs 65 million year ago. What is not generally recognized is how much is at stake. Few people realized, for example, that 40% of the world economy is derived directly from biodiversity. At present, natural habitats and ecosystem are being destroyed at the rate of over 100 million hectares ever year. Some 9 million hectares of forests are lost ever year; 50% of the world's wetlands were lost in the past century and 80% of grasslands are suffering from soil degradation. Some 100-150 species are lost every day. If the current destruction rate in forests and coral reefs is maintained, 50% of the Earth's plants and animal species will be gone by the end of the twenty-first century.

Second only to habitat destruction, invasive alien species represent a major threat to biodiversity, with consequential economic loss. One study documents that the economic damage associated with such species in six selected countries total more than US$ 336 billion per year and assumes that similar costs of damage worldwide from invasive species would be more than US$ 1.5 trillion per year. Anthropogenic changes in the atmosphere also threaten to accelerate global rates of extinction. According to one scenario, if the current rate of global warming continues, one- third of the planet's species will disappear by 2050.

The underlying causes of biodiversity loss are diverse and complex. They include:

- increasing demand for biological resources as a result of increasing population, economic development and over-consumption.

- failure of people to appreciate the consequences of using inappropriate technology.

- failure of economic markets to recognize the true value of biodiversity at local levels.

- failure of government policies to recognize and address the problems associated with the over-use of biological resources.

- increasing human migration, travel and international trade.

3. History

A General Agreement on Tariffs and Trade (GATT) was established in 1947 (came into force on 1 January 1948) to deal with post-world war scenario concerning multilateral trade issues, particularly with a view to serve as means of stimulating world trade in goods, remove/ minimize impediments to trade among member countries, and to create a level playing field. In Uruguay the 8th round of such GATT negotiations started in 1986 but it was concluded on April 15, 1994 at Marrakesh, Morocco with the signing of an accord by 123 countries that led to the formation of World Trade Organization (WTO) on 1st January 1995. This round was significant because agriculture was included as a tradable commodity and agreement on TRIPs was achieved.

The International Undertaking on Plant Genetic Resources (IUPGR) that had been advocating free exchange of plant genetic resources (PGRs) for research use, being the heritage of mankind, released a revised agreed interpretation in 1991 that such free access will be subject to the sovereign rights of nations over their PGRs. This is directly relevant to both the CBD and the TRIPs Agreement.

Convention on Biological Diversity (CBD) was conceived during the UN Conference on Environment and Development at the Earth Summit in Rio de Janeiro in 1992 with respect to biological resources. It deals with the sovereign rights of the nations over their genetic resources. CBD came into force in December 1993. The CBD Convention has more than 170 parties. It aims to secure the conservation and sustainable use of biological diversity. This is the first legal mechanism which has very categorically highlighted that plant genetic resources are not to be the heritage of humankind, but are the properties of the countries and hence these are tradable commodities.

4. Treaties, Conventions and Undertakings

4.1 The International Undertaking on Plant Genetic Resources (the "Undertaking")

Food and Agriculture Organization (FAO) has helped to generate several non-legally binding international instruments relating to plant genetic resources. The Undertaking, the first of these instruments, was adopted in 1989 that advocated free exchange of plant genetic resources (PGR) for research use, being the 'heritage of mankind' (Resolution No. 4/89). It simultaneously recognized Farmers' Rights as defined in Resolution 5/89. An interpretation clarified that plant breeders' rights were not incompatible with the Undertaking. It also recognized the interrelationship between the rights of traditional farmers (whose practice of saving seeds provided the raw genetic material for innovation) and the rights of plant breeders (who use technology to achieve that innovation). IUPGR that had been advocating free exchange of plant genetic resources (PGRs) for research use, being the heritage of mankind, released a revised agreed interpretation in 1991 that such free access will be subject to the sovereign rights of nations over their PGRs (Resolution 3/91, thus pledging themselves to implement the recommendations it contains. For many years, the Undertaking served as the central legal instrument in FAO's global system for plant genetic resources, a system that includes a fund for the equitable sharing of benefits and a mechanism to give early warning about genetic resources under threat. The Undertaking's principal objectives are to ensure that the need for conservation is globally recognized and that sufficient funds for this purpose are made available; to assist farmers and farming communities in the protection and conservation of PGRs and of the natural biosphere; and to allow farmers, their communities and countries to participate fully in the benefits derived from improved uses of PGRs, including through plant breeding. This is directly relevant to both the CBD and the TRIPs Agreement.

4.2 *International Treaty on Plant Genetic Resources for Food and Agriculture ("ITPGRFA") – Seed Treaty*

On 3rd November 2001, an intergovernmental conference sponsored by FAO adopted the text of a legally binding international agreement on plant genetic resources through resolution 3/2001. These negotiations culminated in the adoption of the *International Treaty on Plant Genetic Resources for Food and Agriculture* (ITPGRFA) also known as Seed Treaty. The ITPGR entered into force on 29 June 2004. The treaty had been ratified by 55 states and signed by an additional 50 countries. The ITPGR not only codifies and updates the nonbinding principles set out in the Undertaking and its subsequent revisions, but also contains provisions relevant to IPRs in plant genetic resources and plant varieties. The IU prescribes a multilateral system of access and benefit sharing for genetic resources for food and agriculture. It is based on the premise that the free exchange of genetic resources is essential to food security and the conservation of genetic resources for food and agriculture. The Treaty recognized the enormous contribution that the local and indigenous communities and farmers have made and will continue the efforts on conservation and development of plant genetic resources. It recognized following rights as main components of the farmers' right, (i) Right to save, use, exchange and sell farm-saved seed and other propagating material, (ii) Right to fair and equitable sharing of the benefits arising from the use of plant genetic resources for food and agriculture, (iii) Right to participate in national decision-making process about plant genetic resources [entrusting national governments with the responsibility for implementing these rights in accordance with their needs and priorities subject to national legislation], and (iv) Protection of traditional knowledge. Thus, a full-fledged concept of farmers' rights came to an existence at the international forum.

5. Convention on Biological Diversity

In June 1992, the United Nations Conference on Environment and Development (UNCED) was held in Rio de Janeiro, Brazil. It was also called the 'Earth Summit'. One of the major outcomes of UNCED was Agenda 21 and another was the Convention on Biological Diversity (CBD), a new international law that is set to have a major impact on the way we deal with life forms. CBD came into force in December 1993. At the UNCED meeting 153 countries signed the new agreement, many of whom have subsequently ratified thereby accepting its provisions and agreeing to work to promote its implementation. By 2015, 168 countries had ratified it and 196 parties are signatory to it.

CBD defines Biological diversity as "variability among living organisms from all sources and the ecological complexes of which they are part and includes diversity within species or between species and of ecosystems". The objectives of the CBD set out a balance between conservation, sustainable use and benefit-sharing (Article1). CBD defines biological resource as "plants, animals and micro-organisms or parts thereof, their genetic material and byproducts with actual or potential use or value but does not include human genetic material". CBD defines 'genetic resources' as 'genetic material of actual or potential value. 'Genetic material' is any material of plant, animal, microbial or other origin containing functional units of heredity. Documentation of a genetic resource can also cover its parts or components, such as organs, cells, cell organelles, genes, etc.

In order to fulfill the obligations of CBD, nations were bestowed with sovereign rights over biological resources within their territories. The access to these biological resources is to be regulated by national legislation. However, any grant of access to these resources must fulfill the condition of "prior and informed consent of the relevant stakeholders under mutually agreed terms".

The objectives of CBD are:

i. the conservation of biological diversity;

ii. the sustainable use of the components of biological diversity; and

iii. the equitable sharing of the benefits derived from the use of genetic resources; including ensuring: relevant access to genetic resources; the transfer of relevant technologies; that appropriate funding is available.

One of the ways in which CBD differs from other international conventions is that it sets goals, rather than any specific targets or objective for the countries ('parties') seeking to implement it. There are no lists of priority habitats, sites or species to be conserved. The CBD leaves it up to individual national governments to decide how it is to be implemented. Measuring biodiversity or its conservation is a pretty impossible task but nevertheless, the CBD has provided an extremely important internationally accepted frame work within which countries can operate and cooperate in implementing broadly similar actions for biodiversity conservation. At the heart of the CBD are provisions on scientific and technical cooperation, access to genetic resources, technology transfers and a financial mechanism (the Global Environment Facility) to help implement the CBD in developing countries.

The contracting parties are also given ample power and discretion to implement national measures to achieve the objectives. The Convention provide general measures to be adopted for conservation and sustainable use of biological diversity and fair and equitable benefit sharing arrangements between providers and users of relevant resources. These rights and broad discretions vested in the nation are not absolute but are subject to an obligation scheme for biodiversity. CBD has completely changed the characteristic of genetic resources from "common heritage of mankind" to "sovereign property of the countries".

Some of the features of the convention that can affect the global IPR regime are as follows:

Sovereign rights of state: Sovereign rights of states over their natural resources. The authority to determine accesses to genetic resources rests on the national governments and is subject to national legislation [Art 15.1].

Recognition of the contribution of local population: Recognition of the contribution of local population to respect, preserve and maintain the knowledge, innovation and practices of indigenous and local communities relevant to biodiversity conservation and utilization [Art 8j]. It further promotes a wider application of such traditional knowledge with the approval and involvement of holders and to encourage the equitable sharing of the benefits arising from the utilization of such knowledge.

Access to genetic resources: The convention has recognized the role of national governments in determining access to genetic resources within their national territory on mutually agreed terms and prior informed consent for providing such access [Art 15.3].

Access to transfer of technology: It has stressed the adoption of legislative, administrative and policy measures so as to provide access to transfer of technology to developing countries on mutually agreed terms, including technology protected by patents and other IPRs [Art 16.3].

Sharing of results and benefits: The contracting parties to the CBD are obliged to take legislative, administrative or policy measures, with the aim of sharing in a fair and equitable way, the results of technology and development and benefits arising from the commercial and other utilization of genetic resources with the contracting party providing such resources [Art 15.7].

One of the mechanisms by which the CBD achieves its objectives is *in situ* conservation that involves the preservation of ecosystems and natural habitats and the maintenance of viable population of species in those settings. Such conservation occurs, for example, where farmers and indigenous communities safeguard traditional plant varieties in the locations where they grow naturally or are cultivated.

Although the CBD does not expressly refer to any international IPR agreements, it contains numerous provisions relating to IPRs, principally in article 16. In particular, article 16(5) recognizes that IPRs "may have an influence on the implementation" of the CBD. The article obliges member states to cooperate in order to ensure that IPRs are "supportive of and do not run counter to" the treaty's objectives. Articles 16(2),(4) state that the transfer of technology and measures taken to gain access to such technology shall be consistent with the adequate and effective protection of IPRs recognized in international law. Thus, for example, where a government encourages foreign direct investment in industrial technologies (such as a biotechnological process used to insert new genetic sequences into existing plant varieties), it must respect any patent rights that the owner of that technology has acquired to protect it.

Over the time, CBD has constituted the Conference of the Parties (COP) to streamline and implement the CBD directives. It has given detailed attention to harmonizing IPRs with the CBD's objectives. In particular, developing countries active in the COP, together with the support of nongovernmental organizations, have expressed concern about the adverse effects of IPRs and have sought to harness intellectual property rules to promote compliance with the Convention.

In response to this concern, official COP statements have stressed the need to "promot[e] increased mutual supportiveness and integration of biological diversity concerns and the protection of intellectual property rights." (CBD Decision IV/15, para. 9). In April 2002, the COP adopted the "**Bonn Guidelines** on Access to Genetic Resources and Fair and Equitable Sharing of the Benefits arising out of their Utilization." (CBD Decision VI/24, App. II, Annex, Part C). The Bonn guidelines provides the parties and stakeholders with a transparent framework to facilitate access to genetic resources and ensure fair and equitable sharing of benefits through standard practices and procedures of prior informed consent (PIC), mutually agreed terms (MAT), material transfer agreement (MTA) and other relevant agreements. The guidelines provide details of an overall strategy and the essential steps, elements and principles to be adopted in developing access and benefit sharing (ABS) regime by the parties and stakeholders. The Guidelines' most important recommendation encourages applicants for IPRs such as patents or plant breeders rights to disclose the country of origin of the genetic resources or the traditional knowledge upon which those IPRs are based. The Guidelines advocate these disclosures to monitor whether applicants for IPRs have obtained the prior informed consent of the country of origin and complied with the conditions of access (if any) that country has adopted.

6. Indian Biological Diversity Act 2002

The key development in India concerning biodiversity is the enactment of legislation on The Biological Diversity Act, 2002 and Biological Diversity Rules, 2004. The Act was passed on 5 February 2003 pursuant to Article 253 of the Indian Constitution. A National Biodiversity Authority has been created under the Ministry of Environment and Forests, with its headquarters at Chennai. The main principle enshrined in the Indian Biological Diversity Act, 2002 is: "...to provide for conservation of Biological Diversity, sustainable use of its components and equitable sharing of the benefits arising out of use of biological resources and for matters connected therewith or incidental thereto." These aims are in line with the objectives mentioned in the CBD.

The Act defines the terms "biological diversity" and "biological resources". "Biological diversity" refers to the variability among living organisms from all sources and the ecological complexes of which they are part and includes diversity within species or between species and of ecosystems. "Biological resources" means plants, animals and microorganisms or parts thereof, their genetic material and by-products (excluding value-added products) with actual or potential use or value but does not include human genetic material. Thus, one can see that human genetic material has been expressly excluded from the scope of the Act, while ago-biodiversity is brought within its realm.

The grant of intellectual property rights related to plant varieties shall be outside the authority of the National Biodiversity Authority (NBA) since it has been exclusively regulated by the Protection of Plant Varieties and Farmers' Rights Act, 2001.

The Act makes substantial provisions in the following aspects:

- Access to biological resources.
- Research activity.
- Effect on Intellectual Property Law
- Benefit-sharing
- Penalties for contravention of law.

6.1 Access to Biological Resources

In the Act it has been provided that foreigners who intend to carry out such activities as research, commercial utilization, bio-survey or bio-utilization relating to biological resources occurring in India, or knowledge associated thereto, shall seek prior approval of the NBA by filing an application to that effect. The categories of such foreigners who need approval have been defined in Sec. 3 of the Act as:

a) a person who is not a citizen of India,

b) a citizen of India who is a non-resident as defined in Clause (30) of Sec. 2 of the Income Tax Act, 1961, or

c) a corporate body, association or organization which is not incorporated or registered in India, or incorporated or registered in India under any law for the time being in force which has any non-Indian participation in its share capital or management.

To the contrary, Indian scientists or organizations can perform bio-surveys and bio-utilization or commercial utilization of biological resources by simply giving prior intimation to the State Biodiversity Boards (SBB) under Sec 7 of the Act. They are not required to wait for the approval of the SBB and can undertake these activities forthwith, unless a formal decision of rejection is sent to them.

6.2 Research Activity

Difference exists for Indian nationals and foreigners regarding research activities involving biological resources. Research carried out by Indian persons or organizations on biological resources are exempted even from the condition of prior intimation to the SBB. However, no person shall, without the previous approval of NBA, transfer the results of any research relating to any biological resources occurring in, or obtained from India for monetary consideration or otherwise to any foreigner as defined above under Sec 3 of the Act. Transfer does not include publication of research papers or dissemination of knowledge in any seminar or workshop. However, this provision shall not apply to collaborative research projects involving transfer or exchange of biological resources or information relating thereto between Government institutions and such institutions in other countries, if such collaborative research projects satisfy the conditions of approval by the policy guidelines of Govt. of India.

6.3 Effect on Intellectual Property Law

The BDA Act under Sec. 6 clearly lays down that no person (this includes both Indian nationals and foreigners) shall apply for any intellectual property right in or outside India for any invention based on research or information on biological resources obtained from India without the approval of the NBA. This provision is in consonance with the long-standing Indian demand of making amendment to the TRIPs Agreement to incorporate requirements of prior informed consent and disclosure of country of origin of biological resources. The Act also requires that permission of NBA is a prerequisite for seeking patent/IPR on any Indian bioresource accessed by foreigners. This is in conformity with CBD, which recognizes that States have sovereign rights over their own biological resources and that the authority to determine access to genetic resources rests with the national governments and is subject to national legislation.

The NBA while granting the approval under this section 6 may impose benefit sharing fee or royalty or both or impose conditions including the sharing of financial benefits arising out of commercial utilization of such rights. However provisions of this section shall not apply to local people and communities of the area including growers and cultivators of biodiversity, and *vaids* and *hakims*, who have been practicing indigenous medicine.

Under Sec. 6 of the BDA Act no person shall apply for any IPR, in or outside India for any invention based on any research or information on a biological resource obtained from India as mentioned above. If a person applies for a patent, permission of NBA may be obtained after the acceptance of the patent but before sealing of the patent. Section 10(D) of the Patents Act also clearly stipulates that where a patent application relates to biological material, the applicant shall disclose the source and geographical origin of such biological material in the specification. Under 64(1)(p) of the Patents Act, a patent can be revoked if the complete specification does not disclose or wrongly mention the source or geographical origin of biological material used for the invention. Furthermore, the Act provides for exclusion of certain inventions from the purview of patentable invention if they are based on traditional knowledge.

Contravention of Sec. 6 of the Act only entails criminal penalties of imprisonment and fines; it does not provide for the revocation of patents. Such revocation is dealt with by the Patents Act. Thus, both laws complement each other in order to provide a comprehensive legal framework to address the issue of biopiracy. However, the grant of patent rights in a foreign territory in contravention of the Act does pose a problem. Being a national law, it only has territorial applicability. In such a scenario, the Government either has to file an application for revocation or oppose the grant of patent, based on the choices available in the foreign jurisdiction.

Provisions relating to intellectual property in the Act apply equally to both national and foreign persons. This is based on the principle of "equality in national treatment" enshrined under Article 3 of TRIPs, that each country must treat nationals of other WTO Members in similar favourable ways as it treats its own nationals. In other words, IPR protection and enforcement must be non discriminatory as to nationality of right holders.

6.4 Benefit Sharing

Sharing of the benefits arising from research, bio-survey or commercial utilization of biological resources carried out by foreigners, as defined under Sec. 3 of the Act, should be determined through an application made to the NBA. This procedure is entirely different from the one that is applicable to Indian nationals and organizations. In the case of Indian nationals, it is the SBB which has the jurisdiction.

The NBA shall ensure, while granting the approval to foreign individuals, that fair and equitable benefit-sharing takes place between the foreigners, the local bodies concerned and the benefit claimers. Benefit sharing can be in any of the following ways:

a. grant of joint ownership of IPRs to NBA or where benefit claimers are identified, to such benefit claimers

b. transfer of technology

c. location of production, R&D units in such areas which will facilitate better living standards to the benefit claimers

d. association of Indian scientists, benefit claimers and the local people with R&D in biological resources and bio-survey and bio-utilization

e. payment of monetary compensation and other non-monetary benefits to the benefit claimers as the NBA my deem fit.

In situations where any amount of money is ordered by way of benefit sharing, then NBA may direct the amount to be deposited in National Biodiversity Fund.

However, no such direction has been given to the SBB for Indian nationals. This situation arises because Indian nationals are only required to submit prior intimation and not to seek any approval for undertaking their activities. They can begin their research activities based on the knowledge and resources of local communities forthwith, without waiting for any approval from the SBB.

Sec. 23 of the BDA Act states that the function of the SBB shall be, inter alia, to regulate the granting of approvals or otherwise requests for commercial utilization, bio-surveys and bio-utilization of any biological resource by Indian nationals and to perform such other functions as may be necessary to carry out the provisions of the Act or as may be prescribed by the state government.

Thus, access to biological resources to foreigners has been given but regulated by NBA, while Indian companies and research organizations are given freedom in terms of access and commercial exploitation. In the case of the latter, it is only when a product is made a subject-matter of intellectual property rights that the principle of benefit-sharing will be applicable, which shall then be monitored and supervised by the NBA.

6.5   Penalties for Contravention of the Law

The sanctions imposed upon persons and organizations of India origin and on foreigners as defined under Sec. 3(2) and Sec. 4 regarding research activity and Sec. 6 IPR on biological resources shall be punishable with imprisonment for a term up to 5 years, or with fine up to ten lakh rupees and where the damage caused exceeds ten lakh rupees such fine may commensurate with the damage caused, or with both [Sec. 55 (1)]. However for Indian nationals or a body corporate, association or organization registered in India contravenes by obtaining any biological resource for commercial utilization under Sec. 7 shall be punishable with imprisonment for a term which may extend to three years, or with fine which may extend to 5 lakh rupees, or with both [Sec. 55 (2)].

6.6   A Decentralized System of Management, Implementation and Enforcement

A decentralized system has been created based at national, state and local levels with constitution of the NBA, the SBB and the Biodiversity Management Committees (BMCs). The salient features of this three-tier structure is the involvement of self-governing institutions of panchayats and municipalities in ensuring the conservation and sustainable use of biological resources.

The National Biodiversity Authority: NBA is vested with powers to regulate activities mentioned in Sections 3,4 and 6 of the Act and to issue guidelines for access to biological resources, grant approvals, regulation for fair and equitable benefit-sharing. It advise the central government on matters relating to the conservation of biodiversity, sustainable use of its components and equitable sharing of benefits arising out of the utilization of biological resources. It also advise the State Governments in the selection of areas of biodiversity importance to be notified under Sec. 37(1) of the Act as heritage sites and measures for the management of such heritage sites;

The State Biodiversity Boards: State Biodiversity Boards are to be created by the state Governments through official notification and are primarily responsible for scrutinizing the prior intimations received from Indian nationals and organizations for commercial utilization of biological resources. The constitution and the administrative and financial composition of the SBB is the same as that of the NBA.

The Biodiversity Management Committee: At the grass roots level are the Biodiversity Management Committees (BMC), which shall be constituted by local bodies. "Local body" is defined to mean panchayats and municipalities. Under Sec. 41 of the Act function of BMCs is to promote conservation, sustainable use and documentation of biological diversity including preservation of habitats, conservation of land races, folk varieties and cultivars, domesticated stocks and breeds of animals and microorganisms and chronicling of knowledge relating to biological diversity. For this purpose terms have been clearly defined:

Cultivar means a variety of plant that has originated and persisted under cultivation or was specifically bred for the purpose of cultivation

Folk variety means a cultivated variety of plant that was developed, grown and exchanged informally among farmers

Landrace means primitive cultivar that was grown by ancient farmers and their successors.

Coordination or cooperation between BMCs, SBB and the NBA is sought to be achieved between for use of biological resources and knowledge associated with them and occurring within the territorial jurisdiction of the BMCs.

6.7 Traditional Knowledge in the Biodiversity Act

There is no separate mention of "protection of traditional knowledge" in the Act, but terms such as "knowledge" or "information" are used obliquely to refer to the concept of traditional knowledge. This "knowledge" or "information" which is sought to be protected by the Act is qualified by another criterion: that it should relate to or be associated with biological resources. Some measures are incorporated within the Act to ensure monetary compensation to the providers of knowledge where the commercial exploitation of biological resources or knowledge was a result of access given by a specific individual or group of individuals.

Under Sec. 36(5) of the Act, it is provided that the Central Government shall endeavour to respect and protect the knowledge of local people relating to biological diversity through such measures as recommended by the NBA. That could include registration of the knowledge and / or creation of a *sui generis* system for protecting such knowledge.

6.8 Conservation and Sustainable Use of Biological Resources

Appropriate duties, function and authorities have been assigned to the NBA, the SBB and the BMCs for conservation and sustainable use of biological resources. Apart from this, under Sec. 9 of the Act, the Central Government has been assigned the duties of developing national strategies, plans etc. for conservation and sustainable use of biological diversity, including measures for identification and monitoring areas rich in biological resources, promotion of *in situ* and *ex situ* conservation of biodiversity, incentives for research, training and public education to increase awareness with respect to biodiversity. The Government shall also undertake measures for assessment of the environmental impact of projects and will regulate, manage or control the risk associated with the use and release of living modified organisms. State government under Sec. 37 of the Act, have been given power to notify areas of biodiversity importance in the Official Gazette as biodiversity heritage sites.

7.   Case studies on Benefit Sharing

There are very few examples or models of benefit sharing arising out of the use of genetic resources from the holders of Traditional knowledge and Traditional medicine. India has the distinction of being the first in the world in experimenting a benefit-sharing model that implemented Article 8(j) of Biological Diversity Act in letter & spirit.

7. 1   TBGRI Model or Kani Model

It relates to medicine development which, is based on the active ingredient of Arogyapacha *(Trichopus zeylanicus)* plant. Scientists at Tropical Botanic Gardens and Research Institute (TBGRI), Thiruvananthapuram, Kerala, India were on an expedition to areas inhabited by kani tribals in 1987. Their interaction with the tribals led them to an observation of a plant that the tribals chewed to keep themselves energetic. They studied the properties of this plant by isolating and testing the active ingredients of the plant and incorporated these ingredients into a product, which they christened 'Jeevani' on the basis of knowledge of Kani tribals. This product was shown to bolster the immune system. It was later found that it had immuno modulation hepato-protection as well aphrodisiac properties. The development of this drug was entirely possible due to the indigenous knowledge of the Kani tribe residing in the Western Ghat forests in India. TBGRI, which is the sole patent holder struck an agreement with the tribal community to share 50% license fee and continues to receive the royalty at the rate of 50% of royalty that TBGRI gets from the party. To transfer the benefits the tribals were encouraged to form a trust, Kerala Kani Samudaya Kshema Trust (KKSKT). As per rules of the trust the license fee and royalty received on account of the sale of jeevani drug will be in a fixed deposit and only the interest accrued will be utilized for the benefits. In addition to the license fee and royalty, a large number of Kani families are now getting benefits from the cultivation of Arogyapacha and supply of raw material to the pharmaceutical company for the production of drug. This marks perhaps for the first time that compensation in the form of cash benefits has gone directly to the Kani tribals, traditional knowledge holders.

7.2 Hoodia (Hoodia gordonii)

For thousands of years, African tribesmen have eaten the Hoodia cactus to stave off hunger and thirst on long hunting trips. The kung bushmen, San, who live around the Kalahari desert in Southern Africa used to cut off a stem of the cactus about the size of a cucumber and munch on it over a couple of days. In 1995, the South African Council of Scientific and Industrial Research (SACSIR) patented Hoodia's appetite-suppressing element (P57) and hence, its potential cure for obesity. In 1997, they licensed P57 to British Biotech Company, Phytopharm. In 1998, Pfizer acquired the rights to develop and market P57 as a potential slimming drug and cure for obesity (a market worth more than £ 6 billion), from Phytopharm for $ 32 million. The San people eventually learned of this exploitation of their traditional knowledge, and in June 2001, launched legal action against the SACSIR and the pharmaceutical industry on ground of biopiracy. They claimed that their traditional knowledge had been stolen, and the SACSIR had failed to comply with the rules of CBD, which require the PIC of all stakeholders, including the original discoverers and users.

The two sides entered into negotiation for a benefit-sharing agreement, despite complication regarding as to who should be compensated, that is, the persons who originally shared the information, their descendants, the tribe, or the entire country. The San are nomads spread across four countries. In March 2002, a landmark agreement was reached according to which the San will receive a share of any future royalties. The settlement will not directly affect Phytopharm or Pfizer since the San would be paid out of the royalties that would be received by the SACSIR as the latter was the patent holder. The South African Council will probably receive a royalty of around 10% from Phytopharm, which itself will receive royalties from sales by Pfizer. Thus, San are likely to end up with only a very small percentage of eventual sales.

7.3 Xa21 gene

One of the most serious bacterial diseases of rice is bacterial blight caused by *Xanthomonas oryzae* pv. *oryzae*. Dr Davadath found genetic source of resistance to this disease in a species of *Oryza longistaminata* (wild rice from Mali). The blight resistant specimen was transferred to International Rice Research Institute (IRRI), Manila, Philippines for breeding purposes in 1978. Scientists at IRRI namely Dr G.S. Khush, Dr R. Ikeda and coworkers introduced the resistance found in the above mentioned sample from India in to cultivated varieties using conventional plant breeding methods. They discovered that the resistance was contributed by a single locus called Xa21. At Cornell University this gene was mapped. From 1992-1995 high resolution mapping. DNA library construction, cloning and sequencing was carried out at University of California, Davis, USA leading to isolation of a few candidate clones carrying Xa21 gene. Collaboration with Lili Chen at International Laboratory for Tropical Agricultural Biology in La Jolla, CA, USA co-directed by C. Fauquet and R. Beachy transformed a susceptible Taipei 309 rice variety with Xa21 carrying clones. The resulting plants assayed at Davis, USA showed high level of resistance to bacterial blight. A patent application No 475,891 entitled "Nucleic acids, from *Oryza sativa*, which encode leucine rich repeat polypeptides and enhance Xanthomonas resistance in plants" was filed at USPTO. Patent was granted on Jan 12, 1999. This patent on *Xa*21 gene became a pivotal tool for benefit sharing arrangements. In June 1996 UC Davis established Genetic Resource Recognition Fund (GRRF) to recognize the contributions of various developing countries to the success of *Xa*21 cloning. The GRRF shall be used for providing fellowships to the students from Mali, Philippines and other developing countries. For whatever reasons, the companies which had licensed the gene for commercial exploitation either did not utilize it or generate any commercial returns from its application. Hence, no funds had yet been received by GRRF for benefit sharing.

7.4. Bioresource Development and Conservation Programme in the Field of Traditional Medicine in Nigeria

It relates to IPRs (patents, trademark and copyright) acquired over value added biological resources and associated knowledge which are expected to generate profits from the traditional knowledge of Nigerian people. This case study focuses on two patent applications filed for inventions related to the work of Shaman Pharmaceuticals and International Co-operative Biodiversity Group (ICBG) programme. On Dec., 19, 1989, Shaman Pharmaceuticals Inc. filed a patent application (No. 452,902), granted on May 28, 1991 with US Patent No. 501,580 entitled "Dioscoretine and its use as a hypoglycemic agent". Inventor in the application is Prof. Maurice M. Iwu. On Dec 18, 1990 Shaman Pharmaceuticals Inc filed an international patent application under PCT with international publication No. WO 91/09018 with priority data of application No. 452,902, dated 19.12.89 and inventor Prof. Maurice M. Iwu. Inventions relate to a novel biologically active compound, more particularly dioscoretine, isolated originally from tubers of *Dioscorea dumetorum*. Tubers were collected at Ankpa local government area in the Benue state of Nigeria. Novel isolated compound dioscoretine is useful as a hypoglycemic agent and thus provides a new and useful agent for the treatment of diabetes mellitus. Patents cited above were granted prior to entry into force of CBD. Thus, these were governed by the research agreement with Shaman which provided for a benefit sharing plan. A share of the royalties generated from commercialization of the technology and licensing will be contributed by Shaman Pharmaceuticals Inc to the trust fund established for benefit sharing with traditional healers and local communities in Nigeria and other countries that have been working with the ICBG programme. Shaman followed the approach of obtaining the clear prior informed consent of various countries and institutions. Shaman pioneered a concept of short term and long term benefits to the local communities. Shaman's compensation included upfront payment to the local people or host country capacity building. Technology and resource transfer for capacity building was initiated in 1990, when collaboration began, even no commercializable product was developed from Nigerian plants. An amount of USD 210,000 was provided to Nigerian stakeholders which has enabled significant capacity building at BDCP, augmentation of Phytotherapy Research Lab at Nsukka, traditional healer organization and the rural communities. However, the company has made a huge net loss. Shaman has not come out yet with any marketable products nor have they shared the benefits with the indigenous or local people or any other stakeholder.

These are the examples where mechanisms of returning some of the benefits from the commercialization of medicinal plants and traditional knowledge, to the local people have been developed.

8. Establishment of the Traditional Knowledge Digital Library

Grant of patents on non-original innovations, which are based on what is already a part of the traditional knowledge in developing world, has been causing a great concern in these countries. India has taken a positive view on IPR regime that patents will be granted by different Patent Offices of the world, but patents should not be granted wrongly on TK, TM or on genetic resources. To counteract the biopiracy, India has established a Traditional Knowledge Digital Library (TKDL) related to Ayurveda, Unani, Yoga and Sidha in five international languages, viz., English, French, Spanish, German and Japanese. It is a collaborative project between National Institute of Science Communication and Information Resources (NISCAIR), Ministry of Science & Technology, Department of Ayurveda, Yoga and Naturopathy, Unani, Sidha and Homeopathy, Ministry of Health and Family Welfare; and the Office of Controller General of Patents, Designs and Trade Marks, Department of Industrial Policy & Promotion, Ministry of Commerce & Industry.

TKDL is a digital database of traditional knowledge consisting of about 250,000 formulations in the field of TM, with a view to prevent patenting of such knowledge and thus avoid misinterpretation of publicly available information as being an invention or a discovery. The project documents the knowledge available in public domain by sifting and collating the information on traditional knowledge from the existing literature in patent application format. It covers Ayurveda, Yoga, Siddha and Unani systems of traditional medicines.

Since Traditional Knowledge documentation lacked classification, Trasitional Knowledge Resource Clssification (TKRC), an innovative structured classification system for the purpose of systematic arrangement, dissemination and retrieval has been evolved for about 8000 sub groups against one group in International Patent Classification (IPC). Access to TKDL is to be provided free but under a Non Disclosure Agreement (TKDL Access Agreement), due to the sensitivity of the database to different patent Offices of the world. TKDL will act as bridge between traditional knowledge existing local languages and patent examiners at International Patent Offices and it is expected to prevent the misappropriation of India's rich Traditional Knowledge.

Selected Reading

http//www.biodiv.org/doc, UNEP, Convention on Biological Diversity, Rio de Janerio

http//www.wto.org

http//www.wipo.org

http//www.nbaindia.org

Pushpangadan, P. 2010. Access and benefit sharing- Indian experience. In: Compendium on Intellectual Property Rights and Development: A national Perspective (Eds Yeshodharan, E.P.), Kerala State CST&E, Thiruvananthapuram, Kerala, pp 23-38

Traditional Knowledge

Traditional knowledge (TK) is considered as the modern term which is directly linked with tradition or culture of respective countries of the world. In short, TK is a community based functional knowledge system developed, preserved and refined by generations of people through continuous interactions, observations, experimentation with the surrounding environment. It is a dynamic system ever changing, adapting and adjusting to the local and religious practices of the communities. TK also encompasses the wisdom, knowledge, teaching and experience of these communities and usually it is orally transmitted from generation to generation. The definition of TK is widely debated and discussed and it is yet to arrive at consensus and hence a universal definition could not be provided because of its nature, characteristics, uses, etc. Traditional knowledge (TK), Indigenous Knowledge (IK), Traditional Environmental Knowledge (TEW) and Local Knowledge (LK) are relative terms coined by researchers, policy makers or scholars belonging to different traditions from time to time according to its origin and utility of genetic resources which is associated with biodiversity.

The products based on traditional knowledge are important sources of income, food and healthcare for large parts of the population in developing countries in particular, and, in turn, for their sustainable socio-economic development. What makes traditional knowledge 'traditional' is not its antiquity, but the way it is acquired and used. In other words, the social process of learning and sharing knowledge, which is unique to each culture, lies at the very heart of its traditionality.

WIPO currently uses the term traditional knowledge to refer to tradition based literary, artistic or scientific works; performances; inventions; scientific discoveries; designs; marks; names; symbols; undisclosed information; and all other tradition - based innovations and creations resulting from intellectual activity in the scientific, industrial, literary or artistic fields. Tradition – based refers to knowledge systems, creations, innovations and cultural expressions which have generally been transmitted from generation to generation; they are generally regarded as pertaining to particular people or its territory; and are constantly evolving in response to changing environment. Traditional knowledge functions under two streams: 1) Classical tradition (codified) that have a vast published literature; and 2) oral tradition (not codified) which is prevalent in rural and tribal villages of a country.

Western societies, in general, had not recognized any significant value of traditional knowledge. They had also not recognized the obligations associated with its use. These societies, in context of their standard intellectual property laws, also looked at traditional knowledge as information in the 'public domain', which was freely available for use by anybody. The concept of any compensation to the creators and possessors of traditional knowledge also did not exist. It is only recently that western science has become more interested in traditional knowledge. They are beginning to realize that traditional knowledge, in combination with modern scientific knowledge, can lead to the solution of current problems in diverse areas ranging from agriculture to health.

1. Traditional Knowledge Protection: Strategies

The issue of protection of traditional knowledge needs to be looked at from two perspectives. The protection may be granted to exclude the unauthorized use by third parties of the protected traditional knowledge and the protection may also mean the preservation of traditional knowledge from uses that may erode it or negatively affect the life or culture of the communities that have developed and applied it.

One of the concerns of the developing world is that the process of globalization sometimes seen as appropriating the elements of the collective knowledge of societies into proprietary knowledge for the commercial benefit of a few. Action is needed to protect these knowledge systems through national policies and international understanding community knowledge and community innovation.

The local communities or individuals do not have the knowledge or the means to safeguard their property in a system which has its origin in very different cultural values and attitudes. The communities have a storehouse of knowledge about their flora and fauna-their habits, their habitats, their seasonal behavior and the like-and it is only logical and in consonance with natural justice that they are given a greater say, as a matter of right. A policy that does not obstruct the advancement of knowledge, and provides for valid and sustainable use and adequate intellectual property protection with just benefit sharing, is what the world needs. Discussion of intellectual property rights and traditional knowledge should draw more on the diversity and creativity of indigenous approaches to IPR issues.

Several proposals have been made, within and outside the IPR system, to protect traditional knowledge (TK). The possibility of applying the existing modes of IPRs protection to different components of TK has been extensively explored. Some elements of traditional medicine may be protected under patents. However, since most of the TK is not contemporary and has been used for long periods, the novelty and/ or inventive step requirements of patent protection may be difficult to meet. It would be easier to comply with a more flexible novelty requirement such as that for plant varieties in UPOV that had been previously commercialized or disposed of for purposes of exploitation. Some valuable TK may be kept secret, such as in cases of applications of plants for therapeutic purposes. Holders of this knowledge may be protected against disclosure under unfair competition rules. Geographical indications may, in some cases, be a suitable mechanism to enhance the value of agriculture products, handicrafts and other TK-derived products. Copyright can be used to protect the artistic manifestation of TK holders, especially artists who belong to indigenous and native communities, against unauthorized reproduction and exploitation. Another approach would be the development of a *sui generic* regime of IPRs which is specifically adapted to the nature and characteristic of TK. *Sui generis* is a Latin term denotes "of its own kind" is used to describe something that is unique or different. The conventional IPR regimes introduced by different countries/organizations of the world did not provide adequate protection for TK. There are big gaps in many places in terms of policies, programmes, legal framework and guidelines. This gap has to be identified and adequate protective measures are to be framed and implemented at international, regional, national and local level.

Normally debate on IPRs and biodiversity has focused on patents and plant breeders' rights. Provision concerning undisclosed information or trade secrets could be invoked to protect traditional knowledge not available in the public domain. Geographical indications and trademarks, or *sui generic* analogies, could also be the alternative tools for indigenous and local communities seeking to gain economic benefits from their traditional knowledge. The potential value of geographical indications and trademarks is in protecting plants and germplasm that are specific and unique to geographical region. They could protect and reward traditions while allowing innovation.

2. International Initiatives for Protection of Traditional Knowledge

The Convention of Biological Diversity (CBD) is the most appropriate for understanding the Access and Benefit Sharing (ABS). CBD signed at the UNCED in 1992, was the first international environmental Convention to develop measures for the use and protection of Traditional Knowledge (TK), related to the conservation and sustainable use of biodiversity.

2.1 Convention on Biological Diversity

Article 8 (j) encourages National Governments to develop mechanisms for equitable benefit sharing. Each Party to the CBD shall, 'as far as possible and as appropriate':

Respect, preserve and maintain the knowledge, innovations and practices of indigenous and local communities embodying traditional lifestyles relevant for the conservation and sustainable use of biodiversity, promote their wider application with the approval and involvement of the holders of such knowledge, innovations and practices; Encourage the equitable sharing of the benefits arising for the utilization of their knowledge, innovations and practices.

Article 10: Sustainable use of Components of Biological Diversity. Each contracting party shall, as far as possible and as appropriate.

Article 10 C: Protect and encourage customary use of biological resources in accordance with traditional cultural practices that are compatible with conservation or sustainable use requirements. (The interpretation of these provisions has been elaborated through decision by the parties (rectifiers of the convention).

Article 15: It sets out a frame work to achieve the objective. It recognized the sovereign rights of respective country over their natural resources and suggests the National agencies/Authorities set the conditionality for determining access to genetic resources.

2.2 The Bonn Guidelines

The Bonn Guidelines repeatedly propose 'support measures' and 'capacity building' as a requirement in order to address the evident imbalance between parties.

2.3 WTO & WIPO Initiatives

In November 2001, a ministerial conference of WTO members agreed the Doha Mandate in Qatar. This mandate identified areas for further negotiations, one of them being the relationship between TRIPs and CBD's protection of Traditional Knowledge. WIPO has under taken work on ABS related issues since 2001, though the Inter-Governmental committee Intellectual Property Rights and Genetic Resources, Traditional Knowledge and Folklore.

3. Intellectual Property Rights and Traditional Knowledge

Two extreme views have surfaced in the debates in the CBD and others on protection of traditional knowledge:

  i. One view advocates the extension of intellectual property protection to cover traditional knowledge.

  ii. The second view promotes the *status quo* where such knowledge is treated as a public good.

Those who are in favour of the first view argue that extending intellectual property protection to traditional knowledge will in fact promote technological innovation, as it would facilitate the dissemination and development of that knowledge in the modern economic space. An example of how the intellectual property system can be utilized is the case of Aboriginal and Torres Strait Islander artists in Australia. They have obtained a national certification trademark which is intended to promote the marketing of their art and culture products and deter the sale of products falsely claimed to be of aboriginal origin. Further, recognition of intellectual property rights in traditional knowledge could generate incentives amongst the indigenous communities in conserving and preserving their environment.

Also, the industrialized countries have a moral obligation to ensure that indigenous and local people receive a fair and equitable share of benefits arising from the use of their traditional knowledge and commercialization of their genetic resources. The opportunity of exploiting traditional knowledge comes from interlinking and developing partnerships between local innovators and entrepreneurs on the one hand and by holder of traditional knowledge on the other.

The Honeybee Network in India, for example, works to protect the intellectual property rights of grassroots level innovators through the documentation and dissemination of their innovation via the said Network, and has compiled a very large database. This activity is based on the fundamental belief that when people's knowledge is collected and recorded, they should not become poorer for sharing their insights and for connecting innovators through the networking in local language. If income is generated by developing people's knowledge, they must be rewarded with a fair share. Another example is in the People's Democratic Republic of Laos, where the government established the Traditional Medicines Resource Center (TMRC), which is working with the local healers to document details of all traditional medicines with a view to promoting a practice of mutually beneficial sharing. The TMRC is also collaborating with the International Cooperative Biodiversity Group (ICBG) in efforts to discover prospective medicinal products. Any benefits, profits or royalties realized from plants and knowledge recovered during the collaboration will be shared with all the involved communities.

Opponents of the extension of intellectual property protection to traditional knowledge argue that such a move might destroy the social basis for generating and managing the knowledge being community property, passed on from one generation to the next, it would be privatized through protection under intellectual property law, and this may deny access to such knowledge.

Intellectual property rights provide an incentive to turn knowledge into a marketable commodity and to commercialize it. Mere preserving of such knowledge in its cultural context, although laudable, does not serve a wider public good. This provides a potential to use these for a wider possible public good, rather than confining them locally. Indeed traditional knowledge has been increasingly used for providing the 'technical lead' in biodiversity prospecting. A number of pharmaceutical companies rely on traditional knowledge of indigenous and local people in their screening activities for identifying the biologically active constituents which have potential commercial market. While doing so, the rights of these people have been ignored. Indigenous and local people have not been shared in a fair and equitable manner, the benefits arising from appropriation of their knowledge and its subsequent use in drug development. Though it is fully recognized that there is a need to protect traditional knowledge, and to secure a fair and equitable sharing derived from the use of biodiversity and associated traditional knowledge, there is no agreement *per se* on what would be the most appropriate and effective way of achieving these objectives.

4.   Protection of Traditional Knowledge Documentation

Traditional knowledge documentation data constitute an important form of non-patent literature with specific characteristics making it a category of non-original database. Some of those characteristics may necessitate specialized measures for traditional knowledge data to be adequately integrated and recognized as relevant non-patent literature and as a non-original database.

If the development of such measures takes into account the needs and priorities of all stakeholders, they might: (i) avoid the grant by IP offices of patents for traditional knowledge-based inventions which are not novel and non-obvious; (ii) avoid the costs of challenging such patents for traditional knowledge holders and other interested third parties; and (iii) facilitate recognition of the technological value of traditional knowledge by all users of non-patent literature, including IP offices, industry, researchers and the general public.

The grant of patents on non-original innovations (linked to TM), which are either based on what is already a part of the traditional knowledge or a minor variation thereof, have been causing a great concern to the developing world. A recent study by an Indian expert group examined randomly selected 762 US patents, which were granted under A61K35/78 and other IPC classes, having a direct relationship with medicinal plants in terms of their full text. Out of these 762 patents, 374 (49%) were found to be based on traditional knowledge. The IPC Union has decided to include about 200 sub-group on medicinal plants in contrast to one sub-group A61K 35/17 available at present. In the 32nd IPC Union meeting held in WIPO, Geneva, in February 2003, it was decided that the Task Force should continue its work on further development of classification tools for TK and to investigate possible patent classification aspects relating to components of biodiversity and folklore. The Task Force is also supposed to consider how the future revised IPC could be linked to TKRC, which may be developed in various countries, and how to best organize access to TK documentation, which was in the public domain, including hyperlinking the IPC to TK databases. These developments augur well for integrating traditional knowledge into knowledge system based on industrial system.

Traditional Knowledge Resource Classification (TKRC) devised by India is a classification system for the purpose of systematic arrangement, dissemination and retrieval of traditional knowledge resourc es. It is expected to facilitate the digitization of traditional knowledge and act as a meta library to provide a language- independent storage and retrieval of digitized information. This is devised by following the internationally well-accepted IPC structure which includes sections, classes, sub-classes, group and sub-group.

Patent examiners, in such patent offices, when considering the patentability of any claimed subject matter, use available resources for searching the appropriate non-patent literature sources. Patent literature, however, is usually wholly contained in several distinctive databases and can be more easily searched and retrieved. A traditional knowledge digital library (TKDL) would help to prevent patenting products based on traditional knowledge.

Grant of patents on non-original innovations, which are based on what is already a part of the traditional knowledge in developing world, has been causing a great concern in these countries. India has taken a positive view on IPR regime that patents will be granted by different Patent Offices of the world, but patents should not be granted wrongly on TK, TM or on genetic resources. To counteract the biopiracy, India has established a Traditional Knowledge Digital Library (TKDL) related to Ayurveda, Unani, Yoga and Sidha in five international languages, viz., English, French, Spanish, German and Japanese. It is a collaborative project between National Institute of Science Communication and Information Resources (NISCAIR), Ministry of Science & Technology, Department of Ayurveda, Yoga and Naturopathy, Unani, Sidha and Homeopathy, Ministry of Health and Family Welfare; and the Office of Controller General of Patents, Designs and Trade Marks, Department of Industrial Policy & Promotion, Ministry of Commerce & Industry. TKDL is a digital database of traditional knowledge consisting of about 2.08 lakh formulations in the field of TM, with a view to prevent patenting of such knowledge and thus avoid misinterpretation of publicly available information as being an invention or a discovery. The project documents the knowledge available in public domain by sifting and collating the information on traditional knowledge from the existing literature in patent application format. It covers Ayurveda, Yoga, Siddha and Unani systems of traditional medicines. Each Sloka has been transcribed in approximately four pages. Therefore, the portal will contain approximately 1,40,000 pages in a single language. One time cost on TKDL on Ayurveda has been estimated as US$ 300,000.

India has already shared the TKDL with European Patent Office (EPO) and USPTO on an agreement that EPO and USPTO shall utilize the database for search and examination only and shall not make any third party disclosure except for giving a printout to the inventor/applicant as citation.

Identification of priority medicinal plants was done by the experts of the Department of India System of Medicine and Homeopathy (ISM&H) and CSIR with the help of international patent databases.

Once the National Traditional Knowledge Digital Libraries get created, they may be included in the official list of International Search Authorities (ISA) relating to non-patent literature. Presently, there are 135 non-patent technical journals, in the lists of ISA. Once the TKDL by various member states get created and integrated with www.tkdl.com, it will be useful for IP offices to review patents granted on non-original inventions which are part of the traditional knowledge system. Revocation of such patents by offices will go a long way in addressing the emotive concerns of the developing world on the issue of IPR based on indigenous knowledge.

5. Biopiracy of Traditional Knowledge

The grant of patents on non-original innovations (particularly those linked to traditional medicines), which are based on what is already a part of the traditional knowledge in the same developing world, has been causing great concern in these countries. Grant of wrong patents on knowledge that was already in the public domain has had two effects. Firstly, it has caused heartburn amongst those who possessed this knowledge. Secondly, the challenges to such patents, and their subsequent revocation, have brought more clarity in this debate by identifying the real reasons for grant of such wrong patents and have, on a positive side, led to new initiatives, such as the creation of traditional knowledge digital libraries, and potential changes in the international patent classification system. In the following, some of these cases have been described.

5.1 Challenge to Patents Based on Traditional Knowledge – Decided in Favour of TK Holders

Turmeric (*Curcuma longa*)
The rhizomes of turmeric are used as a spice for flavouring in Indian cooking. Turmeric also has properties that make it an effective ingredient in medicines, cosmetics and as a colour dye. As a medicine, it is traditionally used to heal wounds and rashes. In 1995, two expatriate Indians at the University of Mississippi Medical Centre (Suman K. Das and Hari Har P. Cohly) were granted a US patent (No. 5,401,5041) on the use of turmeric in wound healing. The Council of Scientific & Industrial Research (CSIR) in India, filed a re-examination case with the United States Patent and Trademark Office (USPTO) challenging the patent on the basis of *prior art*. The CSIR argued that turmeric has been used for thousands of years for healing wounds and rashes and, therefore, its medicinal use was not a novel invention. Their claim was supported by documentary evidence of traditional knowledge, including a paper published in 1953 in the Journal of the Indian Medical Association despite an appeal by the patent holders, the Patent Office upheld the objections filed by the CSIR and revoked the patent in 1997, after ascertaining that there was no novelty, the findings reported by innovators having been known in India for centuries. The turmeric case was a landmark judgement as it was the first time that a patent based on the traditional knowledge of a developing country was successfully challenged.

Neem (*Azadirachta indica*)

Neem extracts can be used against hundreds of pests and fungal diseases that attack food crops; the oil extracted from its seeds can be used to cure cold and flu; and mixed in soap, it offers cheap relief from malaria, skin diseases and even meningitis. In 1994, the European Patent Office (EPO) granted a patent (EPO patent No. 436257) to the US Corporation, W.R. Grace Company and US Department of Agriculture for a method of controlling fungi on plants through hydrophobically extracted neem oil. In 1995, a group of international NGOs and representatives of Indian farmers filed a legal petition against the patent. They submitted evidence that the fungicidal effect of extracts of neem seeds had been known and used for centuries in Indian agriculture to protect crops, and thus was a *prior art* and unpatentable. In 1999, the EPO determined that according to the evidence all features of the present claim had been disclosed to the public prior to the patent application and thus the patent was not considered to involve an inventive step. The patent granted on neem was revoked by the EPO in May 2000.

Basmati

Rice Tec. Inc. had applied for registration of a mark "Texmati" before the UK Trademark Registry. It was successfully opposed by the Agricultural and Processed Food Exports Authority (APEDA), India. One of the documents relied upon by Rice Tec as evidence in support of the Rice. Tec. Inc. was US patent 5,663,484 granted by the US Patent Office on September 2, 1997, and that is how this patent became an issue for contest.

This US utility patent (No. 5,663,484) claimed a rice plant having characteristics similar to the traditional Indian Basmati rice lines, and with the geographical delimitation covering North, Central or South America, or the Caribbean Island. The patent had 20 claims covering not only the novel rice plant but also various rice lines; resulting plants and grains, seed deposit claims, method for selecting a rice plant for breeding and propagation. Its claims 15-17 were for a rice grain having characteristics similar to those of Indian Basmati rice lines. The said claims 15-17 would have come in the way of Indian export to the U.S., if legally enforced.

Evidence from the Indian Agriculture Research Institute (IARI), New Delhi, Bulletin was used against claims 15-17. The evidence was backed up by the germplasm collection of the Directorate of Rice Research, Hyderabad, since 1978. The various grain characteristics were evaluated by Central Food Technological Research Institute (CFTRI) scientists and accordingly the claims 15-17 were opposed on the basis of the declarations submitted by CFTRI scientists on grain characteristics. The claims were:

Claim 15: A rice grain which has

i. starch index of about 27 to 35

ii. 2 acetyl-L-pyrroline content of about 150 ppb to about 2000 ppb

iii. length of about 6.2 mm to about 8.0 mm, a width of about 1.6 mm to about 1.9 mm and a length/width ratio of about 3.5 to about 4.5

   iv. a whole grain index of about 41 to 63

   v. a lengthwise increase of about 75% to about 150% when cooked and

   vi. a chalk (starch?) index of less than about 20

Claim 16: The rice grain of claim 15, which has a 2 acetyl-L-pyrroline content of about 350 ppb to about 600 ppb

Claim 17: The rice grain of claim 15, which has a burst index of about 4 to about 1

Eventually, a request for re- examination of this patent was filed on April 28, 2000. Soon after filing the re-examining request, Rice Tec chose to withdraw claims 15-17 along with claim 4. With this patent rights exists only on three lines and their grain *viz.* Bas 867, RT1117 and RT1121. Thus, the patent does not affect any basmati export to USA.

5.2 Challenge to Patents Based on Traditional Knowledge – Decided Against TK Holders/Remain Unresolved

Kava (*Piper mythesticum*)
Kava is an important cash crop in the pacific, where it is highly valued as the source of ceremonial beverage of the same name. Over 100 varieties of kava are grown in the Pacific, especially in Fiji and Vanuatu, where it was first domesticated thousands of years ago. In North America and Europe, kava is now promoted for a variety of uses. The French company L'Oreal has patented the use of kava to reduce hair loss and stimulate hair growth.

Ayahuasca (*Banisteriopsis caapi*)
For generations, shamans of indigenous tribes throughout the Amazon basin have processed the bark of *B. caapi* to produce a ceremonial drink known as "ayahuasca". The shamans use ayahuasca (which means "wine of the soul") in religious and healing ceremonies to diagnose and treat illness, meet with sprits, and divine the future.

   An American national, Loren Miller, obtained a US Plant Patent (No.5,751 issued in 1986), granting him rights over an alleged variety of *B. caapi* which he had collected from a domestic garden in Amazon and had called "Da Vine," and was analyzing it for potential medicinal properties. The patent claimed that Da Vine represented a new and distinct variety of *B. cappi*, primarily because of the flower colour. The Coordinating Body of Indigenous organization of the Amazon Basin (COICA), which represents more than 400 indigenous tribes in the Amazon region, along with others, protested against the wrong patent that was given on a plant species called *B. cappi*. They protested that ayahuasca had been known to be native to the Amazon rainforest, and cultivated for generations for its traditional medicinal uses, so Loren Miler could not have discovered it, and should not have been granted such rights, which in effect, appropriated indigenous traditional knowledge. On re-examination, the USPTO revoked this patent on 3rd November 1999. However, the inventor (Loren Miller) was able to convince the USPTO on 17th

April, 2001, and the original claims were re-confirmed and the patent rights restored to the innovator.

Quinoa (*Chenopodium quinoa*)
One traditional quinoa variety, Apelawa, is the subject of the US patent No.5,304 a staple food crop for millions in the Andes, especially Quechua and Aymara people A patent was held by two professors from the Colorado State University who claim the variety's male sterile cytoplasm is key to developing hybrid quinoa.

To cite some more examples, the plant *Phyllanthus amarus* is used for ayurvedic treatment for jaundice. A US patent has been taken for its use against hepatitis B. The plant *Piper nigrum* is used for Ayurvedic treatment for vitiligo (a skin pigmentation disorder). A patent has been granted in the UK for the application of a molecule from *Piper nigrum* for use in treatment of vitiligo.

In the 1970s the US National Cancer Institute (NCI) invested in large scale collection of *Maytenus buchananii* from Simba Hills of Kenya. The lead for the collection came from the knowledge of the Digo communities-indigenous of the Simba Hills area-who use the plant to treat cancerous conditions. The plant contains a biologically active constituent maytansine which is considered a potential treatment for pancreatic cancer. The collected material was traded without the consent of the Digo community. Their knowledge of the plant and its medicinal properties was also not rewarded. Similarly, the NCI also collected another plant *Homalanthus nutans* from the Samoa rainforests, which contains an anti-HIV compound prostratin. This collection was undertaken on the basis of traditional knowledge. NCI has also benefited from traditional knowledge of local communities living around Korup Forest Reserve in Cameroon, from where it collected *Ancistrocladus korrupensis* to screen for an anti-HIV chemical, Michellamine B. The NCI and other drug research and development organizations have started investing considerable sums of money to prospect for plants containing useful chemicals, and are also investigating the efficacy of traditional medicines.

5.3 Successful Cases of Benefit Sharing

In contrast to the above two categories, where patents were either revoked or not revoked, but the holders of knowledge did not get involved in any financial transaction, there are cases, where the originators of knowledge did get a benefit. Such interesting cases are discussed below.

1. **TBGRI Model or Kani Model** (See Chapter 12 on Biodiversity)

2. **Hoodia (*Hoodia gordonii*)** (See Chapter 12 on Biodiversity)

3. **Xa21 gene** (See Chapter 12 on Biodiversity)

4. **Bioresource Development and Conservation Programme in the field of Traditional Medicine in Nigeria** (See Chapter 12 on Biodiversity)

6. Traditional Resource Rights (TRR)

Knowledge of traditional resources is central to the maintenance of identity of indigenous and local communities embodying traditional lifestyles. Therefore, control over these resources is of central concern in their struggle over land and territory. The term Traditional Resources Rights (TRR) has emerged to define the many "bundles of rights" that can be used for protection, compensation, and conservation. The change in terminology from IPR to TRR reflects an attempt to build on the concept of IPR protection and compensation, while recognizing that traditional resources-both tangible and intangible-are also covered under a significant number of international agreements that can be used to form the basis for a *sui generic* system. "Traditional resources" include plants, animals, and other material objects that may have sacred, ceremonial, heritage, or aesthetic qualities. "Property" for indigenous peoples and local communities frequently has intangible, spiritual manifestations, and, although worthy of protection, can belong to no human being. Indigenous and traditional communities are increasingly involved in market economies, and are seeing an ever-growing number of their resources traded in those markets. Even so, for many, privatization or commoditization of their resources is not only foreign, but incomprehensible or even unthinkable.

The TRR concept can be implemented by identifying guiding principles for legislative processes, and by forming the basis for practical instruments and mechanisms that guarantee protection, benefit sharing, and political and financial support for indigenous and local communities. Governments now have a unique opportunity to support work in the development of a TRR-type *sui generic* system. By doing so they will be taking an important step towards synergizing and harmonizing their human rights commitments with biodiversity conservation, sustainable development and global trade agreements. Governments could assist these processes by providing resources and expertise for the development of practical instruments (such as Material Transfer Agreements, Information Transfer Agreements, contracts and covenants) and mechanisms that embody TRR principles. Financial and political support for indigenous and local communities can also be provided through new guidelines for policies and projects.

Selected Reading

Pushpangadan, P. 2010. Access and benefit sharing- Indian experience. *In*: Compendium on Intellectual Property Rights and Development: A national Perspective (Eds. Yeshodharan, E.P.), Kerala State CST&E, Thiruvananthapuram, Kerala, pp. 23-38

Rajasekharan, S. 2010. Protection of traditional knowledge and *Sui generis* system. *In*: Compendium on Intellectual Property Rights and Development: A national Perspective (Eds Yeshodharan, E.P.), Kerala State CST&E, Thiruvananthapuram, Kerala, pp. 79-92

http//www.wipo.org

PART 3

··

IPR Protection:
Biosafety and Ethical Issues

Biosafety

After the advent of rDNA technology in 1970s, discussions began within the scientific community about the risks associated with recombinant DNA/ genetic engineering experiments. The main concerns were: i) Molecular biologists may not be well-versed with the laboratory practices needed for such type of work; ii) 'Hybrid organisms' could be created with biological activities of an unpredictable nature and has the potential for infection of workers; iii) 'Hybrid organisms' may escape from the laboratory with unpredictable consequences since their survival and behavior in the environments would be largely unknown. National Academy of Sciences, USA in 1974 examined the various issues and made certain recommendations and also established the Recombinant Advisory Committee. In 1975, at Asilomer, California, an International meeting was held which formulated the first set of recommendations on safety of recombinant DNA experiments. These formed the basis of subsequent biosafety guidelines and regulations in USA and other countries. Initially the regulations on recombinant DNA and the genetic modification of organisms focused on the technology itself. Therefore, biosafety concerns were primarily focused towards the safety procedures for recombinant DNA work within the laboratory. The emphasis was to ensure that researchers take proper steps to contain organisms that potentially posed a risk to themselves or human health generally.

Biosafety refers to policies and procedures adopted to ensure environmental safety during the course of development and commercialization of genetically modified organisms (GMOs). As per CBD, Biosafety is a term used to describe efforts to reduce and eliminate the potential risks resulting from biotechnology and its products

1. Risk Analysis

Risk is defined as "the probability of harm". The risk associated with any action depends on the three elements of the following equation.

Risk = hazard × probability × consequences

A hazard is anything that might conceivably go wrong. A hazard does not in itself constitute a risk. The probability associated with a hazard also depends, in part on the management strategy used to control it. Risk can be under estimated if some hazards are not identified and properly characterized, if the probability of the hazard occurring is greater than expected or if its consequences are more severe than expected.

Risk analysis consists of three steps i.e. risk assessment, risk management and risk communication.

1.1   Risk Assessment

The concept of risk assessment is quite complex and relies on both science and judgment. It involves determination of potential and anticipated effects of rDNA research to concerned workers and of the products of research on human health and environment if they are released deliberately or accidentally in the environment or as a result of their consumption. Risk assessment should be carried out in a scientifically manner and be in accordance with recognized assessment techniques. A lack of scientific knowledge should not be interpreted as an absence of risk or an acceptable risk, but further information may be sought. During laboratory research work risk can be assessed as i) initial risk assessment and ii) comprehensive risk assessment. Initial risk assessment is made by the investigator on the basis of the risk group (RG) to which organism, on which he proposes to conduct experiments. Organisms are classified in to four risk groups (1-4) on the basis of their potential effects on a healthy human adult. After the initial risk assessment, a comprehensive risk assessment should be made where an appropriate level of containment for an experiment should be decided. This assessment should consider organism factors such as virulence, pathogenicity, infection dose, environmental stability, spread, etc. and type of manipulations planned.

Risk assessment procedures may vary from case to case but for GMOs it generally covers the following:

1. Characterization of the host, donor and gene transfer process including the molecular and phenotypic characterization.

2. The modified organism is analysed for the risks of pathogenicity, allergenicity, teratogenicity, etc.

3. Substantial equivalence: Since rDNA technology creates vastly different organisms, it has been felt that safety assessment methodologies may not be sufficient. Thus a concept of substantial equivalence has been promoted, which means that if GMO is substantially equivalent to an existing organism, safety assessment of the product should be made in the context of existing organism. The rationale of substantial equivalence helps in adoption of specific testing procedures. A case by case approach is used in assessment which is very useful particularly in the development of transgenic plants and their products.

4. Effects related to gene transfer and marker genes: Transfer of genetic information may take place by one or more mechanisms such as transformation of released genetic material, transduction (viral transfer of DNA into bacteria) or conjugation by cell to cell contact and exchange of plasmid. Risk assessment also looks into the possibilities of the same, although, the probability of occurring any of the above is extremely low. Further, it should also be assessed whether such gene transfer will confer some selective advantage to new host such as virulence, adherence, substrate utilization or production of bacterial antibiotics. There have been some concerns about the use of antibiotic resistance genes as markers but the same has been largely discounted by conducting extensive risk assessment.

5. Ecological effects: When organism is released into the environment risk assessment is totally different from that for contained use. Almost all countries have developed a series of questions regarding the organism, vectors and procedures and the site into which it is to be released. An example of such questions in case of a transgenic crop is as follows:

    i. What parts of the engineered plant will contain the new protein?

    ii. What non-target species will be exposed to it?

    iii. What is its toxicity to those species?

    iv. What is their expected level of exposure?

    v. What are the likely biological effects of exposure?

1.2 Risk Management

It is the use or application of procedures and means to reduce the negative consequences of a risk to an acceptable level. If GMO is to be produced then identification of DNA sequences encoding the desired trait, choice of marker gene, regulatory sequences for the expression of transgene, transformation method, etc. should be considered. For example, concerns have been expressed about the use of antibiotic resistant gene a marker gene, although several studies unanimously concluded that the risk was immeasurably low. Therefore, ongoing efforts are undergoing to identify other types of genes useful as markers and to develop methods for removing marker genes before GM products are commercialized. In addition, a number of specific promoter gene sequences have been identified, which can turn on gene expression in specific tissue/situations. For example, a leaf specific promoter, directing toxin production in the leaves but not roots, stems or flowers could control a gene encoding a toxin active against a leaf-attacking pest. The method of transformation should be such so as to avoid including any extraneous DNA sequence e.g. direct gene transfer by a gene gun in plants, avoids the potential for inserting unnecessary vector DNA because no vector is used. On the other hand, cells transformed by *Agrobacterium* mediated DNA transfer methods usually contain extra pieces of DNA coming from the vector in addition to the desired gene or genes. Such sequences have generally been proven as safe but it is better to avoid, if possible. It may be noted that right from the initiation of research, risk assessment and management consideration be kept in mind and integrated appropriately into the research plan for production of GMOs.

1.3 Risk Monitoring

In view of the speculations about potential harm from GMOs introduced into the environment, there has been considerable focus on the monitoring to follow the fate of these organisms and the transgenes they carry and to be vigilant about the unanticipated consequences. Monitoring programs have been classified into three categories i.e. experimentation, tracking and surveillance corresponding to progressive scale up in field test and commercialization. In view of different monitoring objectives for each successive stage, there is a need to consider larger geographic sampling areas and longer term observation regimes.

1.4 Risk Communication

Whereas risk assessment and management procedures are intended to identify and minimize potential negative effects on human health and the environment, risk communication is an integral part of biosafety procedures to ensure public acceptance of GMOs. It is important to interact with public at large about the specific risks and actions taken to alleviate them before announcing GMO field tests and commercialization. Insufficient or inaccurate information needs to misperceptions of risk resulting in adverse public opinion. For example, risk controversies like the current debate over GM food can be divided into technical and non-technical components. The technical components are generally regarding the scientific hazards evaluated in a risk assessment and the management options arising from the assessment. The non-technical components include the cultural and ethical issues generally raised by non-experts, allegations about secretive regulatory decisions etc.

It is a valuable exercise to have effective risk communication and some of the communication strategies are as follows:

- Accept and involve the public as a legitimate partner and treat adversaries with respect
- Coordinate, collaborate and provide information through credible sources
- Be honest, frank and open, don't keep secrets and acknowledge mistakes made
- Listen to and acknowledge people's concerns
- Be proactive and speak clearly with a balanced and realistic information strategy
- Meet the needs of the media and identify and train communicators.

Public opinion about biotechnology is based on misperceptions of risk fueled by insufficient or inaccurate information. More fully informed opinions can arise only when people have a better and more realistic understanding of how biotechnology will affect their immediate lives and the environment in which they live. Risk communication is thus an important first step towards public dialogue concerning the development and use of GMOs.

2. Containment

Biological safety in laboratories is achieved by adopting good laboratory practices and containment strategies. The term **Containment** is used in describing the safe methods for managing infectious agents in the laboratory environment where they are being handled or maintained. It is a combination of laboratory procedures, equipment and installations, and host-vector systems designed to minimize accidental release of organisms, their dissemination and survival in the environment and accidental infection of laboratory workers and of persons outside the laboratory. The basic biosafety requirement in the development of a GMO is to limit the spread of GMO and its genetic material. The purpose of containment is to reduce exposure of laboratory workers, other persons, and outside environment of potentially hazardous agents. Containment can be of two types: physical and biological. Containment facilities for different Risk Groups as per the recommendations of World Heath organization (WHO) and as applicable are described.

2.1 Biological Containment (BC)

It is concerned with making changes in GMOs so that hazards are reduced when they are released deliberately or accidentally into the environment. The guidelines require that *E. coli* strains and vectors to be used should be highly debilitated so that they are not able to survive outside the laboratory, transfer the foreign gene into another organism and not able to infect human beings. This can be achieved by the use of: i) auxotrophic mutants of *E. coli*; ii) strains lacking recombination ability (*recA⁻*); iii) plasmid vectors that are non self-transmissible and non-mobilizable; and iv) no transposons or antibiotic resistance genes in *E. coli* strains.

In consideration of biological containment, the vector (plasmid, organelle, or virus) for the recombinant DNA and the host (bacterial, plant or animal cell) in which the vector is propagated in the laboratory will be considered together. Any combination of vector and host which is to provide biological containment must be chosen or constructed to limit the infectivity of vector to specific hosts and control the host-vector survival in the environment.

2.2 Physical Containment (PC)

When there are real physical barriers to prevent the escape of biological where the organism is designed so that it is not able to survive in any environment other than that of laboratory. The objective is to confine recombinant organisms thereby preventing the exposure of the researcher and the environment to the harmful agents. The protection of personnel and the immediate laboratory environment from exposure to infectious agents is provided by good microbiological techniques and the use of appropriate safety equipment, (primary containment). The protection of the environment external to the laboratory from exposure to infectious materials is provided by a combination of facility design and operational practices (secondary containment). There are three elements of containment viz. laboratory practice and technique, safety equipment and facility design. Physical containment was grouped in to four categories which were designated as P1, P2, P3 and P4 but they are now referred to as Biosafety levels

3. Biosafety Levels

These are combination of laboratory practices and techniques, safety equipment and laboratory facilities appropriate for the operations performed and the hazard posed by the infectious agents. The proposed safety levels for work with recombinant DNA technique take into consideration the source of the donor DNA and its disease-producing potential. The four biosafety levels (BL) corresponds to (P1<P2<P3<P4) facilities approximate to 4 risk groups assigned for etiologic agents.

Biosafety Level 1 (BL1): It is appropriate for undergraduate and secondary educational training and teaching laboratories and for other facilities in which work is done with defined and characterized strains of viable microorganisms not known to cause disease in healthy adult human.

Biosafety Level 2: It is applicable in clinical, diagnostic, teaching and other facilities in which work is done with the broad spectrum of indigenous moderate-risk agents. Laboratory workers are required to have specific training in handling pathogenic agents and to be supervised by competent scientists. Accommodation and facilities including safety cabinets are prescribed, especially for handling large volume or high concentrations of agents when aerosols are likely to be created. Access to the laboratory is controlled.

Biosafety level 3: It is applicable to clinical, diagnostic, teaching, research or production facilities in which work in done with indigenous or exotic agents where the potential for infection by aerosols is real and the disease may have serious or lethal consequences. Personnel are required to have specific training in work with these agents and to be supervised by scientists experienced in this kind of work. Specially designed laboratories and precautions including the use of safety cabinets are prescribed and the access is strictly controlled.

Biosafety level 4: It is applicable to work with dangerous and exotic agents which pose a high individual risk of life-threatening disease. Strict training and supervision is required and the work is done in specially designed laboratories under stringent safety conditions, including the use of safety cabinets and positive pressure personnel suits. Access is strictly limited.

4. Good Laboratory Practices

The most important rules are listed below, not necessarily in order of importance:

1. Mouth pipetting should be prohibited.

2. Eating, drinking, smoking, storing food, and applying cosmetics should not be permitted in the laboratory work area.

3. The laboratory should be kept neat, clean and free of materials not pertinent to the work.

4. Work surfaces should be decontaminated at least once a day and after any spill of potentially dangerous material.

5. Members of the staff should wash their hands after handling infectious materials and animals and leaving the laboratory.

6. All technical procedures should be performed in a way that minimizes the creation of aerosols.

7. All contaminated liquid or solid materials should be decontaminated before disposal or reuse; contaminating materials that are to be autoclaved or incinerated at a site away from the laboratory should be placed in the durable leak proof containers, which are closed before being remove from the laboratory.

8. Laboratory coats, gowns, or uniforms should be worn in the laboratory; laboratory clothing should not be worn in non laboratory areas; contaminated clothing should be disinfected by appropriate means.

9. Safety glasses, face shields, or other protective devices should be disinfected by appropriate means.

10. Only persons who have been advised of the potential health hazards and meet any specific entry requirements (immunization) should be allowed to enter the laboratory working areas; laboratory doors would be closed when work is in progress; access to animal houses should be restricted to authorized persons, children are not permitted in laboratory working areas.

11. There should be an insect and rodent control program.

12. Animals not involved in the work being performed should not be permitted in the laboratory.

13. The use of hypodermic needles and syringes should be restricted to parenteral injection and aspiration fluids from laboratory animals and diaphragm vaccine bottles. Hypodermic needles and syringes should not be used as a substitute for automatic pipetting devices in the manipulation of infectious fluids. Cannulas should be used instead of sharp needles wherever possible.

14. Gloves should be worn for all procedures that may involve accidental direct contact with blood, infectious materials, or infected animals. Gloves should be remove aseptically and autoclaved with other laboratory wastes before disposal. When disposable gloves are not available, re-usable gloves should be used. For removal they should be cleaned and disinfected before re-use.

15. All spills, accidents and overt or potential exposure to infectious materials should be reported immediately to the laboratory supervisor. A written record should be prepared and maintained. Appropriate medical evaluation, surveillance, and treatment should be provided.

16. Baseline serum samples may be collected from and stored for all laboratory and other at risk personnel. Additionally serum specimens may be collected periodically depending on the agents handled or the function of the facility.

17. The laboratory supervisor should ensure that training in laboratory safety is provided. A safety or operations manual that identify known and potential hazards and that specifies practices and procedures to minimize or eliminate such risks should be adopted. Personnel should be advised of special hazards and required to read and follow standard practices and procedures.

4.1 Greenhouse Containment

Greenhouse facilities should be made as per the specifications required for the category of transgenic plant being handled. Conventional greenhouses designed to keep insects and animals out and plant and plant parts are in can be made suitable for GMOs by structural upgrades. For higher level of containment facilities have to meet specifications such as controlled and filtered airflow, systems to control and disinfect water leaving the facility, autoclave for on-site sterilization of plant material and equipment, disinfecting the facility after experiments, strict limits on whom is allowed to enter, and staff and worker training.

5. Risk Management in Field Trials

The procedures vary depending on the nature and magnitude of identified risks, which in turn depend on basic characterization of the organism, the nature of genetic modification and most important, the site where the GMO is to be released. The local ecosystem of the site should be carefully examined before planning the release of a GMO. Some of the risk management strategies are explained below:

For preventing and minimizing the unintentional spread of GMO or a genetic material, measures should be taken to confine them within a site/zone having designated borders/limits. This can be used by both physical and biological means.

5.1 Physical Strategies for Confinement

i. Physical means to confine GMOs particularly in case of plants and animals to geographically or spatially isolate by the use of structures such as fences, screens, mesh etc. The access to the site should be controlled.

ii. In case of plants, appropriate isolation distance should be worked out to control the fertilization of sexually compatible species growing in the vicinity by transgenic pollen. It is essential to collect information about the presence and distribution of cross-fertile wild or weedy relatives of cultivated species near the proposed site.

iii. In case when spatial isolation is not possible due to paucity of land, the following procedures are also used to reduce or prevent the spread of GMO or transgene through pollen or seeds.

    a. Border rows of the non-GM variety may be planted around the test plot to "trap" pollen from the GMO.

    b. Flowering structures may be covered to screen out pollinating insects and/or prevent pollen spread by insect vectors, wind, or mechanical transfer.

    c. Female flowers may be covered after pollination to prevent loss or dissemination of GM seed.

    d. In case where research objectives do not require seed production for analysis or subsequent planting, flower heads may be removed before pollen and seed production.

    e. Plant material of experimental interested may be harvested before sexual maturity.

    f. Test plots may be located surrounded by roads or buildings.

5.2 Biological Strategies for Confinement

A common method of biological confinement is reproductive isolation, which can be achieved by adopting some of the following strategies.

i. GM plants may be grown in an area where sexually compatible wild or weedy species are not found.

ii. All plants of sexually compatible wild or weedy species found within the known effective pollinating distance of the GM crop may be removed.

iii. Flowers may be cover with bags to screen out insect pollinators or prevent wind pollination

iv. Production of viable pollen may be prevented by using genetic male sterility, applying a gametocyte, or removing all reproductive structures at an early stage of development.

v. Tubers, rhizomes, storage roots, and all tissues capable of developing into mature plants under natural conditions may be recovered.

vi. Differences in flowering time may be exploited so that GM pollen is not shed at the time when sexually compatible plants nearby are receptive.

6. Other Procedures

i. Incorporating genes into chloroplast DNA instead of chromosomal DNA.

ii. Genetically engineer transgenic plant to produce sterile seeds. This technology was developed as a "technology protection" system to secure intellectual property rights for the improved seed (the so-called Terminator gene). It is highly effective for risk-management purposes, but has raised ethical questions regarding seed saving and the role of multinational corporations in controlling seed and therefore food supplies in developing countries. This technology is not preferred and also not recommended for release of any transgenic plant containing this technology or terminator gene sequences.

iii. Environmental conditions such as temperature, water supply and humidity can also be manipulated to limit reproduction, survival or dissemination of GMOs outside the experimental area.

iv. Chemicals such as herbicides, fungicides, insecticides and disinfectants can also be used to limit survival and reproduction of GMOs outside the trial area.

v. At the end of the experiment, the whole experimental area can be sterilized or treated with appropriate chemicals.

It may be noted that these measures will be applicable on case-by-case basis on organism with novel traits. Generally the risks are acceptably low at the time of field-testing, due to earlier extensive research being conducted during research and development.

6.1 General Precautions

Careful records of all experiments need to be kept as they provide documentation of the genetic modification and verification data, observed phenotype and any other unexpected observations. This information is necessary not only for documenting performance in the field but also to ensure compliance with risk management and redress in the case of accidental release. The monitoring procedures are applied in such a way that appropriate measures can be taken in case of unexpected effects during or after the release.

It is also important to have appropriate termination procedure to ensure that the GMOs are effectively removed from the experimental area. The required measures are determined by the type of organism, their natural means of spread and the environment in which testing was carried out. Some form of disinfecting is necessary for microorganisms whereas harvesting seeds and ploughing in or burning residual plant material is usually effective in case of plants. Detailed guidelines have been developed for the same, which should be followed.

It may be noted that the biosafety risks to human health and environment can be reduced to acceptable levels by careful management. In addition to the science based issues, the risk management options also take into consideration the policies of the regulatory authorities and the measures that are possible scientifically and economically. The costs of risk management are borne by the applicant in essentially every country and therefore it is important to ensure that the necessary risk management requirements are worked out in such a way that both private as well as public funded research organizations can undertake the same.

7. Biosafety Concerns

Conventional breeding involves crossing related species to develop plants with desired characteristics which are selected among the progeny for reproduction and the selection is repeated over many generations. Genetic engineering bypasses reproduction altogether. It transfers gene horizontally from one individual to another as opposed to vertical gene transfer from parent to offspring. It makes use of infectious agents as vectors or carriers of genes so that genes can be transferred between distant species that would never interbreed in nature. Due to horizontal gene transfer in the development of GMOs there have been conflicting reports on the benefits of GMOs, risks, apprehensions, environmental concerns, and social concerns for the release of GMOs. Some researchers argue that transgenic organisms pose potentially no risks than the conventionally modified organisms. Some argue that technologically modified genes cannot be reliably assessed with the present incomplete knowledge about living systems. Research is being conducted on field tests of transgenic plants and data is generated and the potential risks from the use of GMOs and products thereof have been described. Risk, as explained earlier, means different things to different people in different situations. It can be thought of as a combination of probability and consequence i.e. the likelihood of an event multiplied by the impact of the event. Formal risk assessment usually considers: i) what can go wrong? ii) how likely is it to happen? and iii) what are the consequences if it does not happen? It is important to recognize that just because an event may happen, does not mean it will happen. Once the chance of a risk occurring has been calculated, political, social, cultural and economic considerations will determine whether people believe it to be acceptable. Discussions on GMOs have shown that risk perceptions differ dramatically, even between experts, depending on individuals, their motives and values. Broadly these risks have been categorized under risks to human health, environmental concerns and social and ethical grounds.

8.   Risk to Human Health

8.1   Risk of Toxicity

The risk of toxicity may be directly related to the nature of the product whose synthesis is controlled by the transgene or the changes in the metabolism and the composition of the organisms resulting from gene transfer. In some cases organism may contain inactive pathways, with the addition of new genetic material which, could reactivate these inactive pathways or otherwise increase the level of toxic substances within the plants. Further, the modified metabolism due to introduction of tolerance to chemical substances such as herbicides may also lead to appearance of novel metabolites in the cell. For example herbicide resistant varieties have been released to permit the use of glyphosate and bromoxynil herbicides for weed control. It has been reported that bromoxynil causes birth defects in animals and is also toxic to fish. Thus it may cause serious hazards to farm workers. Glyphosate is also reported to be toxic to soil organisms and to fish. It also accumulates in the fruits and tubers since plants cannot degrade it. Many of the substances involved in plant defence are also known to produce toxic effect e.g. lectin encoding genes. Wheat germ agglutinin (WGA) lectin gene is of plant origin and the product is heat stable and resistant to proteolytic digestion. But in rats it causes loss of weight and hypergrowth of small intestine. Thus it points to a very serious observation that although gene is of plant origin, from a food crop but it does not guarantee the food safety of the gene product. Toxic effects of some of the genes are shown in Table 14.1. In view of the above, every GMO needs to be carefully evaluated for toxicity to human and animals. Most of such

Table 14.1: Toxic and allergenic effects of some compounds (Reviewed in Franck Oberaspach and Keller, 1997)

Compound	Role in Plant defense	Toxic effects	Allergenicity
Lectins	Involved in fungus and insect resistance	Weight loss and hyper-growth of small intestine in rats	Allergenic
Proteinase inhibitors	Insect resistance	Pancreas enlargement, hyperplasia and adenoma	Allergenic
Thionins	Antifungal	Toxic on intravenous injection nontoxic orally (insufficient data)	–
Alkaloids[1]	Insect resistance	Toxicity problems[1]	–
Phytoalexins[2]	Fungus and insect resistance	Toxicity problems[2]	–
Virus coat protein	Virus resistance (transgenic)	Most likely nontoxic (toxicity not reported)	–
Cry protein	Insect resistance	Most likely nontoxic (toxicity not reported)	–

[1] A potato variety derived from a cross with *Solanum chalcoense* had up to 650 mg/kg of glycoalkaloids as compared to 20-150 mg/kg of glycoalkaloids in normal varieties. Variety was withdrawn from the market due to acute toxicity to humans.

[2] A conventionally developed insect resistant variety of celery had 8 fold higher psoralen (phytoalexin) content. Psoralen is known to be toxic, mutagenic and carcinogenic.

toxicity risks can be assessed using scientific methods both qualitatively and quantitatively. The controversy of Starlink corn affecting the human health is mentioned in Box 1.

Box 1

The Starlink corn incident

The US Environment Protection Agency in 1998 approved for use as animal feed a corn modified by insertion of the cry 9C gene from Bt encoding for an insecticidal crystal protein endotoxin by Aventis CropScience. The corn was marketed as StarLink™. Because of concern that the protein Cry 9C could be allergenic, it was not approved for human food. In September 2000, a coalition of environmental and food safety groups announced that cry9C DNA fragments had been found in a popular brand of taco shells sold in the United States. In addition, the Cry9C protein was discovered in some non-StarLink™ seed corn and used as human food. As a result, there was a voluntary recall of corn-derived food products in the United States by manufacturing companies, some of whom took steps, such as mandatory testing requirements, to ensure no further contamination of corn used in human food with Starlink corn.

Late in 2000, a further review of the potential allergenicity of Cry9C, and of mechanisms for assessing suspected allergenic reactions to StarLink™ corn concluded that the Cry9C protein had a medium likelihood of providing to be a potential allergen and that seven out of 34 reactions to a meal containing corn products were probably allergic.

The presence of Cry9C protein in seed corn was due to physical contamination, or it can be due to cross pollination from Starlink corn. Thus this incident raised many issues:

i. The difficulties of restricting a GM food for animals or industrial purposes from that of human food when unmodified crop is used for humans.

ii. The difficulty of preventing accidental physical contamination

iii. The difficulty of ensuring adherence to separation requirements to prevent cross pollination of GM modified and unmodified crop.

iv. The need for appropriate labeling and for post market monitoring to identify allergenic reactions rapidly and correctly.

However, Aventis Crop Science voluntarily took many steps that this Starlink corn is not contaminated with corn used by humans and also withdrew this corn from market even for animal use.

8.2 Risk of Allergies

Production of GMOs sometimes includes the introduction of newer proteins from the organisms which have not been consumed as foods and may cause allergy. All known food allergens are proteinaceous in nature, mostly heat stable and resistant to proteolytic digestion but most proteins do not elicit any allergenic effects on humans. It may be noted that allergies can be developed by any individual to even common foods such as egg, milk, fish, soybean, wheat, etc. Some of the proteins involved in plant defence reactions are known food allergens e.g. peanut lectin, soybean Kunitz trypsin inhibitor, wheat germ agglutinin, barley 15 kDa alpha amylase/trypsin inhibitor (Table 14.1).

The consequences of allergenic reactions depend on the quantity of allergen ingested and duration of consumption as reaction becomes more severe with increasing dose. Also, there is no evidence that GM products pose more risk than conventional products. The possibility of transferring allergens with genetic engineering came to light when a methionine–rich 2S storage albumin protein producing gene from the Brazil nut (*Bertholletia exceisa*) was incorporated into soybean to enhance its sulphur containing amino acids in the seed storage protein. The process was experimented by Pioneer Hi-Bred in the USA. The transgenic soybean was meant for use as an animal feed. Pioneer Hi-Bred considered the possibility that nut protein might finds its way into the human nutrition and therefore the tests were commissioned. Unfortunately the Brazil nut 2S gene was found to be allergenic which raised the biosafety concerns. The tests conducted by the scientists on allergens confirmed that consumption of the transgenic soybean could trigger an allergic response in those sensitive subjects who were allergic to Brazil nut. The company, therefore, decided not to release the transgenic soybean for sale. The allergenic results showed that the undesirable properties of the Brazil nut protein were transferred when the gene was moved into a different species. In general, if a gene produces an allergenic protein in one species, it will likely do so in a new species.

However, sometimes a non-immunogenic protein could become immunogenic when it is expressed in another species. Transgenic pea line has been developed with alpha amylase inhibitor gene which gives protection from pea weevil (*Bruchus pisorum*), but animal feeding experiments revealed that it caused an immune response in mice including inflammation of lungs (Prescott et al., 2005). This was not expected from alpha amylase inhibitor gene but it was found that subtle changes occurred when it was expressed in pea. Thus, there is a strong argument in favour of case by case biosafety assessment of transgenic plants regarding allergenicity assessment.

Specific guidelines have been developed by WHO, FAO, US Food and Drug Authority. Based on these guidelines DBT, regulatory agency in India has formulated a set of guidelines.

8.3 Antibiotic Resistance

Current methods of generating transgenic plants employ a "selectable marker gene" which is transferred along with any other gene of interest usually on the same DNA molecule. The presence of a suitable marker is necessary to facilitate the detection of genetically modified plant tissue during development. The most widely used selectable marker is a bacterial gene for neomycin phosphotransferase (*nptII*), besides *aadA* gene for streptomycin and spectinomycin resistance and the *hpt* gene for resistance to hygromycin. The use of these antibiotic resistance markers has raised the concerns that eating foods carrying antibiotic resistance marker would reduce the effectiveness of antibiotic to fight disease when these antibiotics are taken with meals. The antibiotic resistance gene produce enzyme that can degrade antibiotics. Therefore, theoretically if a transgenic tomato with an antibiotic resistance gene is eaten at the same time as an antibiotic, it could destroy the antibiotic in the stomach. This issue was first raised during the approval process of Calgene's Flavr Savr tomato and Ciba-Geigy's Bt corn 176. A possible concern on the transfer of resistance to gut microorganisms and the potential for transfer of resistance to potentially hazardous microorganisms has been raised.

From various studies it has been concluded that these selectable markers basically have no effect and following points have emerged which are summarized below:

1. The transfer of *npt*II gene from plant material to gut microflora is extremely unlikely because there is no evidence that such transfer can occur through horizontal gene transfer (HGT). For plant DNA to be transferred by HGT a whole set of conditions must occur: i) Available DNA from the plant must be free from the cells, of sufficient length and persist long enough for uptake. DNA in dying plant cells is generally rapidly degraded but it can survive in some soils, aquatic environments or the digestive tract of mice long enough to be available for uptake. ii) A bacterial recipient must be in a suitable state for DNA uptake (competent) and a mechanism for uptake needs to be in place. How often this occurs in bacteria in natural surroundings is unknown but competence can be induced in the laboratory? iii) The recipient cell needs to incorporate, maintain and use the incoming DNA. Integration will depend on sequence homology and a gene will only be useful if it can be read by the recipient's cell machinery. Some studies have shown that gene transfer to plant associated fungi has taken place but there is no evidence for stable integration and subsequent inheritance. Moreover, there are numerous barriers in the gut which make this event extremely unlikely to occur and the event, if it were ever to occur, would be unlikely to be maintained in the absence of constant selection pressure for resistance due to the extremely limited use of these particular antibiotics.

   Although, horizontal transfer of DNA does occur under natural circumstances and laboratory conditions, its probability is extremely rare in the acid environment of the human stomach or even outside environment. Recently, it has been demonstrated that horizontal transfer of DNA containing *npt*II gene (a deletion mutation) can occur at a rate of 1×10^{-8} i.e. one transformant per 10 million cells, under strong kanamycin selection pressure. The probability of such transfer would further decrease in natural circumstances.

2. If the event of a transfer occurs by an unknown mechanism from the genome of genetically modified plants or products derived from them

to gut micro flora and this event being maintained, this would not add significantly to the inherently large microbial population of kanamycin and neomycin resistant microbes in the gut of either humans (Nap et al. 1992) or animals (McAllan et al., 1973; Nap et al., 1992).

3. The expression of *npt*II gene in genetically modified plants is controlled by a plant specific promoter which is not expected to function in bacteria. In the unlikely event of transfer of the *npt*II gene and stable propagation of the intact gene fragment in bacteria, the gene is unlikely to be expressed and even less likely that a DNA rearrangement event occurs that will place the functional nptII encoding open reading frame in front of a bacterial promoter.

4. *npt*II gene if expressed in intestinal bacteria, antibiotic therapy would not be compromised, as the co-factors necessary for the enzyme to inactivate kanamycin and neomycin are not present at the required concentration range in the gut. Moreover the NPTII protein would be rapidly degraded in the gut.

5. In veterinary there is limited usage of kanamycin and neomycin.

6. The *npt*II gene occurs ubiquitously in nature and resistance to this class of antibiotics is already widespread. Therefore, in the highly unlikely situation that a transfer event does occur in gut microflora in humans or animals, this would not significantly impact the overall frequency of kanamycin or neomycin resistant bacteria in the gut or rumen.

Therefore, the overall risk is considered to be effectively zero and the therapeutic use of antibiotics in humans or animals will not be impacted by the commercialization of transgenic crop containing antibiotic resistant selectable marker genes. Although, the ideal situation would be to develop strategies to remove a selectable marker from a transgenic plant prior to commercialization.

8.4 Eating Foreign DNA

There have been apprehensions about danger from eating the foreign DNA in GM foods i.e. the pieces of DNA that did not originally occur in that food plant. DNA being present in all living things such as plants, animals, microorganisms and is eaten by human beings with every meal. Most of it is broken down into more basic molecules during the digestion process whereas a small amount that is not broken down is either absorbed into the blood stream or excreted. In an experiment on feeding mice with a harmless detectable DNA sequence, its progress was tracked through the gastrointestinal tract and the body. About 5% of DNA was detectable in small intestines, large intestines and faeces up to eight hours after the meal, 0.05% in the blood stream up to eight hours; very small fragments in liver and spleen up to 18 hours and no foreign DNA after 42 hours. It has been reported that even if foreign DNA finds it way into tissues of an organism, it is destroyed by body's normal defense system. The DNA of the modified crop will usually be processed and broken down by the digestive system in the same way as that of conventionally bred, or otherwise modified crops. So far there is no evidence that DNA from GMOs including transgenic crops is more dangerous to human health than DNA from conventional crops, animals or associated microorganisms that are normally eaten.

According to a FAO/WHO document on safety aspects of GMOs of plant origin, the amount of DNA which is ingested varies widely, but it is estimated to be in the range of 0.1 to 1.0 gram per day. Novel DNA from a GM crop would represent less than 1/250,000 of the total amount consumed. This means that the possibility of the transfer of genes that have been introduced through genetic modification is extremely low.

8.5 Use of Promoters of Virus Origin

Concerns have been expressed regarding the harm to human health by the use promoters of virus origin e.g. 35S promoter of cauliflower mosaic virus (CaMV). This promoter causes cauliflower mosaic disease in several vegetables, such as cauliflower, broccoli, cabbage and canola. It was suspected that CaMV promoter might be harmful if it invades human cells and turns on certain genes. However, a multi step chain of events would have to occur for the CaMV promoter to escape the normal digestive breakdown process, penetrate a cell of the body and insert itself into a human chromosome. Such a probability is extremely low (virtually impossible), as it has been explained earlier that the normal body defense eliminate any stray fragments of foreign DNA that enter into the bloodstream from the digestive tract.

8.6 Changes in Nutritional Level

Concerns about the accidental changes in the nutritional components of transgenic crops while incorporating other traits have been raised. For example isoflavone levels in Roundup-Ready herbicide resistant transgenic soybean showed minor differences in comparison to conventional varieties. A controversy about increase in phytoestrogen levels in herbicide tolerant soybean has been reported in Box 2.

Box 2

The increase in phytoestrogen levels in herbicide tolerant soybeans can cause breast cancer. A letter from SAG, Basler Appell gegen Gentechnologie, Basle Appeal against Genetic engineering was sent to Swiss Confederate Councillor on Feb. 4, 1997 that we fear Roundup Ready soybean produces large quantities of pseudoestrogens (substances that occur in natural environment and influence or mimic the functions of hormones) when it is sprayed with Roundup herbicide. Today it is assumed that estrogen hormones play an important role in the emergence of breast cancer.

SAG report was based on 1988 report which claimed that conventional French beans produces an estrogenically effective isoflavonoid (coumestrol) after the application of glyphosate (a.i. of Roundup). However, investigations revealed no indication that GM soybeans exhibit any raised concentrations of phytoestrogens following treatment with Roundup herbicide.

9. Risk to Environment

9.1 Gene Flow or Dispersal from Transgenics

Genetic traits with selective advantages in agricultural or natural systems are liable to spread beyond the cultivated variety in which they were originally introduced. Accidental cross breeding between GMO plants and traditional varieties through pollen transfer can contaminate the traditional local varieties with GMO genes resulting in the loss of traditional varieties for the farmers. The wind, rain and insect pollinators can contribute to the spread of pollen resulting in the contamination of local varieties through cross pollination/breeding. In Europe, if contamination by the transgenic pollen exceeds the limit of 1%, then the farmer will not be able to get the label of GM free. Creation of new weeds or increasing the problems posed by existing weeds, due to transfer of modified genes by hybridization or cross-pollination with uncultivated plants or wild relatives, has been suggested as the primary risk in releasing transgenic plants.

Table 14.2: The extent of pollen dissemination from transgenic varieties to other varieties of the same crops (Reviewed by Messeguer, 2003)

Transgenic Crop	Gene/trait	Rate of pollen dispersal (as per cent seed of non-transgenic variety expressing the transgene)
Brassica napus	bar/Glufosinate tolerance	i) 1.5% at 1 m; 0.00033% at 47 m ii) 0.15% at 200 m; 0.00038% at 400 m
Zea mays	bar/Glufosinate tolerance	1% at 18 m
Oryza sativa	pat/glufosinate tolerance	0.1% at 1m; 0.1% at 5 m
Gossypium hirsutum	tfdA/2,4-D resistance; Bt/insect resistance	11.2% at 1 m; 0.03% at 50 m

The likelihood of the transgenes conferring herbicide tolerance, pest resistance or disease resistance being transferred by cross-pollination to sexually compatible species, and the possibility of producing persistent weeds or invasive plant populations, have motivated researchers to study pollen movement in various crops and also sexual compatibility among crops and related species (Table 14.2). Inter-varietal gene flow which occurs frequently and inter-specific gene flow which is a much rarer event are by no means limited to transgenic traits. However, transgenic plants are considered as 'special cases' due to two reasons: (a) many transgenic characters correspond to acquisition of one or several foreign genes, for which no equivalent character is present or would have appeared by spontaneous or induced mutation; and

(b) unlike most characters obtained by empirical selection, transgenic traits are dominant and monogenic, and may thus be readily transmittable by out-crossing.

Further, some of the important factors that may influence gene flow are:

i.   Proximity of the transgenic with compatible wild relatives

ii. Sexual compatibility between crop plant and species or weed

iii. Mating system and mode of pollination

iv. Synchronization of flowering of crop and wild relative or weed

v.   Relative fitness of weed-crop hybrid

vi. Mode of seed dispersal

vii. Nature of transgenic character itself

Studies on gene flow from transgenics irrespective of distance have been conducted. Although out-crossing frequency appears to be strongly influenced by the distance of the recipient plant from the test plot but some studies showed that distance is not the barrier. Hybrids between a transgenic *B. napus* expressing an oil quality modification gene and the weedy *B. campestris* were superior to *B. campestris* in terms of seedling establishment. Transgene introgression can occur from *Beta vulgaris* in to their wild relatives in variable frequencies (Table 14.3). Arias and Rieseberg (1994) reported that physical distance alone will be unlikely to prevent gene flow between cultivated and wild population of sunflower. Skogsmyr (1994) on transgenic potatoes (*Solanum tuberosum* cv. Desiree), containing marker systems like NPTII and GUS, indicated that gene dispersal could occur over long distances and to a higher extent than had been previously shown. Ellstrand (1992) using cultivated radish (*Raphanus sativus* L.), which is out-crossing and insect-

Table 14.3: Examples of gene transfer from transgenic crops to their wild relatives
(Reviewed by Messeguer, 2003, and Malik and Saroha, 1999)

Transgenic crops	Gene/trait	Wild relative	Remarks
Brassica napus	Oil quality modification gene	*Brassica campestris*	Hybrids have advantage over B. campestris during seedling establishment
		Wild radish (*Raphanus raphanistrum*)	Hybrids produced with frequency of ~ 5×10^{-4}
Oryza sativa	Herbicide resistance	Red rice (weedy form)	Red rice seeds showing herbicide resistance ranged from i) 0.8 to 19.7% and ii) 0.002 to 0.6% in different studies.
Beta vulgaris		Wild beet	Gene flow occurs; transgenic progeny found> 200 m behind a strip of hemp plants planted for containment
Solanum tuberosum	—	*Solanum dulcamara* and *Solanum nigrum*	No evidence of transgene transfer to wild relatives.

pollinated, showed that some hybrids containing marker gene could be found even 1km away. A report on corn contaminated by GMO genes in Mexico is reported in Box 3.

Box 3

A report on corn contaminated by GMO genes in Mexico

David Quist and Ignacio Chapela in November 2001 reported that transgenic DNA constructs – 35S promoter sequence together with sequences from alcohol dehydrogenase had been found in a number of creole maize varieties in two remote mountain locations in Oaxacan state of Mexico. The report was widely circulated in media, drawing attention to risks of food security and threats to the genetic diversity. CIMMYT conducted research on 28 accessions from Oaxacan landraces for a study on gene flow. Their initial study could not detect the 35S promoter either in historical accessions from its extensive seed bank or in samples collected recently from the field. In a second phase, the CIMMYT researchers took seeds of 15 maize accessions from gene bank and eight of the accessions were from the state of Oaxaca. DNA was amplified using a primer corresponding to CaMV35S promoter, a fragment of DNA found in most commercial transgenic maize and not known to exist naturally in the maize genome. To further ensure that the reactions are working correctly, all DNA samples were amplified using a primer corresponding to a fragment of DNA known to exist naturally in the maize genome (SSR marker phi076). All positive controls amplified correctly while none of the 15 accessions of gene bank amplified the CaMV35S promoter sequence, indicating that that there is no CaMV35S promoter sequence in the accessions. This was followed by another study on seeds of 42 samples of maize varieties collected from the fields of Oaxaca state. All positive controls (SSR marker phi022) amplified correctly while samples from Oaxaca state did not show the presence of CaMV35S sequence.

As is known maize is an out crossing species and the farmer's varieties are not the same as they were two years ago. Work at CIMMYT has shown that creole maize varieties planted by small farmers in Mexico are constantly changing, both as a result of the biology of plant and traditional practices of farmers. Oaxaca, which is the center of maize diversity, a land race is not stable, uniform or distinct like a plant variety. Farmer's deliberately use external sources of seed to maintain vigour. Gene flow is constant and the real question is whether it makes any difference if one of the genes that have flowed is a transgene. Substitution of one variety for another in an agricultural field does not threaten biodiversity but it is the conversion of native and wild land to agriculture in the first place

Nature, the leading scientific journal disowned the above mentioned paper on April 5, 2002 it published in the 2001 about the environmental safety of GM crops. It followed protests by more than 100 leading biologists who spotted mistakes in the research by two American scientists and attacked Nature for giving respectability to inaccurate results. After studying criticisms of the paper and obtaining new information from its authors, Philip Campbell, the Editor of Nature, agreed that it should not have passed its process of peer review. Author of one of the refutation papers Dr Matthew Metz said the original study relied on flawed techniques "The discovery of transgenes fragmenting and promiscuously scattering throughout the genomes would be unprecedented and is not supported by Quist and Chapela's data.

However, studies also showed that transgenics pose little or no risk through gene flow. Field trials of transgenic rapeseed (*Brassica napus*) with GUS, kanamycin and asulam resistance markers by Paul et al. (1995) and studies by Luby and McNicol (1995) on 80 raspberry populations, using a recessive marker gene *spineless(s)*, and an experiment on a tomato line containing the *anthocyaninless* (*an*) genetic marker by Groenewegen et al. (1994), have indicated the following: (i) limited gene dispersal may occur following large-scale gene deployment; (ii) gene flow events are probably infrequent and appear to be strongly influenced by genotype of the immediately adjacent plants; and (iii) spread is localized for genes having probable selective neutral value. McPartlan and Dale (1994) provided evidence that the extent of gene dispersal from transgenic to non transgenic potatoes fall markedly with increasing distance, and was negligible at 10 m. There was also no evidence of transgene movement from potato to wild species, *Solanum dulcamara* and *S. nigrum*, under field conditions.

Thus, a safe policy would be not to grow transgenic crops where wild relatives occur in proximity of cultivated fields.

9.1.1 Strategies to Prevent Gene Flow of Transgene Escape

Some strategies suggested by researchers to avoid the possible risk of gene dispersal from transgenic plants:

Isolation zone: An isolation zone devoid of vegetation discourages gene flow from transgenic plant to wild or weedy species or to other local cultivars. The isolation distance depends on the mode of reproduction (self or cross-pollination), and also the natural agents promoting cross-pollination (insects/wind, etc.). There are established isolation distances for different crops, mainly used for seed production under controlled conditions which are being followed for experimental studies on gene flow, but their utility in farmers' fields is doubtful.

Trap crop: Use of trap crops, which are non-transgenic varieties of the same crop, planted adjacent to the transgenic plot, can cleanse emigrating pollinators of transgenic pollen, thus preventing gene flow.

Male sterility: Male sterility may be engineered to crops that will not produce pollen or pollen that is inactive, is another suggested mechanism to prevent gene flow from transgenic to sexually compatible species through pollen. Linking of the engineered gene to gene that is lethal in pollen is a mechanism that can provide effective male sterility.

Chloroplast transformation: A highly potential strategy that may be widely applied in future to effectively circumvent the problem of gene flow is **'plastid transformation'** or **'chloroplast transformation'**. In this strategy the gene construct is introduced into chloroplasts and a selection strategy is adopted that allows cells to retain only transformed chloroplasts.

Other strategies: The other proposed strategies are: (i) Removal of flowers from transgenic plants; (ii) Removal of sexually compatible species. But these manual strategies may not find application at a commercial level; iii) Mechanisms like terminator technology and RBF (recoverable block of function) cause embryo lethality and prevent pollen mediated transgene dispersal. Both these techniques are not desirable. As per Indian Protection of Plant Varieties and Farmers Rights Act, 2001 no variety can be registered for commercial exploitation if that contain terminator gene technology or GURT (genetic use restriction technology). **RBF system** consists of a "blocking" sequence, e.g., *barnase* gene, linked to the transgene of interest and a recovering sequence, *e.g.*, *barstar* gene in this case. All the three genes are present in one transformable construct. In tobacco, the *barnase* gene was driven by a germination-specific promoter, while the *barstar* gene was under the control of a heat shock promoter. Under natural conditions, the *barstar* gene is not expressed because the heat shock promoter cannot be induced; as a result, expression of *barnase* in germinating seeds kills the young seedlings.

It is now believed that whenever the crop and weed are different forms of the same species as in sunflower, squash, radish, etc., crop to wild species/weed gene flow can occur when these species happen to grow nearby. Thus, the possibility of gene flow must be carefully examined on a case by case basis.

9.2 Resistance/Tolerance of Target Organisms

The potential benefits of planting insect-resistant transgenic crops include decreased insecticide use and reduced crop damage. However, the innate ability of insect populations to rapidly adapt to environmental pressures poses the development of strains which are resistant to these genes is a serious threat to the long-term efficacy of insect resistant transgenics. Adaptation by insects and other pests to pest protection mechanisms can have environmental and health impacts.

9.3 Generation of New Live Viruses/Super Viruses

Concerns have been raised that due to GM virus resistant plants it could encourage viruses to grow stronger or give rise to new or stronger variants that can infect plants. There are two mechanisms by which this can happen: i) recombination and ii) transcapsidation.

9.3.1 Recombination

In transgenics, viruses recombination can occur either by copy choice or cleavage and ligation. It can occur between GMO produced viral genes and closely related gene of any incoming virus infecting that GMO. Such recombination may produce viruses that can infect a wider range of hosts or that may be more virulent than the parent viruses. This may cause increase in pathogenicity, although there are very few known cases. RNA recombination could also take place between the transcribed transgene and an infecting virus (Greene and Allison, 1994). Evidence about recombination between different viruses has not been collected, but phylogenetic studies of plant RNA virus evolution indicate recombination is common, and some of the new viruses could survive. The frequency of recombination between transgene RNA and viral genomic RNA is unlikely to be higher than that occurring naturally between viral genomic RNAs; and the viability of any new virus is unlikely to be higher than that of existing viruses throughout the infection cycle.

9.3.2 Transcapsidation (Heteroencapsidation)

Transcapsidation is partial or full coating of the nucleic acid of one virus with a coat protein of a differing virus. Thus in a GMO it involves the encapsulation of the genetic material of infecting virus by the GMO produced viral proteins. Such hybrid virus could transfer viral genetic material to a new host plant that it could not otherwise infect. Except in the rare circumstances this would be a one time effect only, because the viral genetic material carries no genes for the foreign proteins within which it was encapsulated and would not be able to produce a second generation of hybrid viruses. So far in the field trials with potato carrying the gene for potato leaf roll virus (PLRV) coat protein and potato viruses S, X and Y failed to acquire PLRV coat protein. Thus it points to the fact that the fears of development of new viruses may not be realistic.

10. Ecological Concerns

10.1 Increased Weediness

The perceived agricultural concerns of GM crops are that these may become weeds or they may invade natural habitats. Although there is no consensus on how to define a weed or the attributes that indicate a plant is likely to become a weed. Weeds do, however, tend to have a preference for disturbed habitats and high physical variability, allowing continuous adaptation to changing environments. Alternatively, a plant without these attributes may become a serious weed if it finds itself in an environment for which it is suited that lacks enemies such as herbivores or diseases. Thus weediness or invasiveness means the tendency of the plant to spread beyond the field where it was first planted, causing undesirable ecological changes. For example, "Kudzu" was introduced into the Southern United State to control soil erosion, but now it has became a major invasive weed in the region. There are apprehensions about GM crops becoming weeds. For example, a salt tolerant GM crop if escapes into marine areas could become a potent weed there.

Many weedy attributes such as dormancy, physical variability, continuous flowering and seeding have been bred out of crop plants over thousands of generations. These traits are now introduced into crop plants using genetic engineering. Some forage grasses, legumes and oilseed rape have a shorter history of domestication and are more likely to become weeds.

One way to assess potential weediness of GM crop is to calculate its finite rate of increase. This calculation is based on the processes affecting population growth, including fecundity, seed survival, seed germination, and seedling survival to maturity. The calculation can take into account any effects of the newly introduced DNA on any step in the process. This analysis has been applied to GM oilseed rape with resistance to the herbicide glufosinate. It has been found that oilseed rape was no more invasive in disturbed or undisturbed habitats than non GM-rape. Calculations of the potential weediness of GM crops, based on population growth studies for oilseed rape and other GM crops showed that these were no more invasive or persistent than their conventional counterparts in a range of environments.

10.2 Creation of Super-weeds

GM crops may hybridize with related weedy relatives and the transgenes (for example herbicide resistance) may be transferred resulting in the creation of superweed. The potential for a crop to hybridize with a weed is highly dependent on sexual compatibility and relatedness. Even if a crop plant and a weed were sufficiently compatible, the survival of any resulting plants would depend on overcoming a number of barriers. Traditional plant breeders use many techniques to encourage plants to hybridize but the process cannot be guaranteed to occur naturally.

One situation where hybridization is more likely to occur is where a crop species is growing alongside its wild relative. In these situations there has always been the potential for crop genes to transfer to the relative. It is important to consider, therefore, whether GM crops are more prone to transferring genes than their non-GM counterparts. For most transgenic traits, GM crops are not likely to transfer either their transgenes, or any other gene, to other species than crop cultivars have done in the past.

Transgenes can spread to wild relatives by cross-pollination, thus creating superweeds. This has occurred in oilseed rape and sugarbeet, creating potential superweeds. Spread of genes by cross-pollination is to be expected, whether the plants are transgenic not. However, a recent report suggests that transgenes may be up to 30 times more likely to escape than the plant's own genes. A large proportion of transgenic varieties under commercial cultivation have herbicide resistance. There is a fear about the development of super-weeds i.e. a weed that acquires the herbicide tolerant gene due to genetic contamination or through horizontal gene transfer.

Environmental organizations believe that weeds will become more resistant through the use of new herbicide resistant transgenic cultivars and their commercialization. Farmers will apply high doses of herbicides over several seasons, rather than limiting their use or using them in rotation with other herbicides, thus increasing chances of development of herbicide resistant weeds. The transgenic plants themselves are already turning up as volunteer plants after the harvest, and have to be controlled by additional sprays of other herbicides. The use of glyphosate with genetically engineered resistant plants will encourage the evolution of glyphosate resistance in weeds and other species, even without cross-pollination. A ryegrass highly resistant to glyphosate has already been found in Australia. Resistance evolves extremely rapidly because all cells have the capability of mutating their genes at high rates to resistance if they are exposed continuously to sub-lethal levels of toxic substances including herbicides, pesticides and antibiotics. This is inherent to the "fluidity" of genes and genomes that has been documented within the past 20 years (Ho, 1998). It will render resistant plants useless after several generations, as the herbicide is widely applied. At the same time, resistant weeds and pathogens may become increasing abundant. Additional herbicides will then have to be used to control the resistant weeds.

Thus there is a need for case by case approach as herbicide resistance may be safe in one location but riskier in another.

10.3 Loss of Biodiversity/Reduction of Cultivars

There have been concerns about reduction in the genetic diversity in cropping systems (i.e. *in situ*) by the development and global spread of improved crop varieties to the green revolution. This genetic erosion has occurred as the farmers have replaced the use of traditional varieties with monocultures. This is expected to further intensify as more and more transgenic crops are introduced which bring in considerable economic benefits to the farmers. The relative rate of susceptibility to any unforeseen infections or destructive situation increases when single varieties are used in cropping system in place of multiple varieties. However, it is argued that there is always a continuous and localized experimentation going on for the development of more effective crops which helps in maintaining genetic diversity. Thus conservation of land races should be done both *in situ* and *ex situ*. Already germplasm banks have been established in various countries.

Herbicide resistant transgenic crops are incompatible with sustainable agriculture. Many studies within the past 10 to 15 years have shown that sustainable organic agriculture can improve yields and regenerate agricultural land degraded by the intensive agriculture of the green revolution. Sustainable organic agriculture depends on maintaining natural soil fertility as well as on mixed cropping and crop rotation. This has been reversing due to the destructive effects of intensive agriculture that have led to falling productivity since 1980s. Glyphosate resistant plants requires application of glyphosate. It is highly toxic to fish and earthworm. It also harms mycorrhizal fungi symbiotically associated with the roots of plants, which are now found to be crucial for maintaining both species diversity and productivity of ecosystems. The depletion of mycorrhizal fungi in intensive agriculture could therefore decrease both plant biodiversity and ecosystem productivity, while increasing ecosystem instability. "The present reduction in biodiversity on Earth and its potential threat to ecosystem stability and sustainability can only be reversed or stopped if whole ecosystems, including ecosystem components other than plants are protected and conserved."

Concerns have been raised that GM plants expressing antimicrobial proteins or Bt toxins could affect soil microbial communities. Bt toxins in soil are estimated to have a half life of 10 to 30 days. The rate of degradation is highly dependent on soil types. Clay particles can bind and inactivate Bt irreversibly. Bt is not taken up and accumulated by other parts and research has shown that microbes near the roots of GM plants are unchanged. The only exceptions reported so far are GM peroxidase producing alfalfa and GM tobacco modified for decreased lignin. It is too early to conclude whether GM crops can have a negative impact on agricultural and natural ecosystems by means of secondary ecological effects. Few examples of secondary effects have been found to date are negative enough to result in problems at an ecosystem level.

10.4 Non-Target Effects

Non-target or unintended effects is another perceived risk as transgenics growing in a particular ecosystem can cause direct or non-target effects on certain microbes or insects growing in a particular ecosystem. GM crops containing insect-resistant Bt (*Bacillus thuringiensis*) toxin have been comprehensively studied. Possible environmental effects include direct effect on non-target insects due to exposure to GM plant material and also any indirect effect on non-target insects via so-called multi-trophic food chains.

For example when Bt cotton was released for commercial cultivation, extensive analysis was carried out on the effect of Bt crystal proteins on various non-target insect populations (honey bees, green lacewing, ladybird beetle, parasitic wasp, etc.), mammals (goat, sheep, buffalo, cow, rabbit, etc), birds and other organisms within the environment. Bt gene product (Cry proteins) is rapidly degraded by the stomach juices of vertebrates, but they could have harmful effects on non-target insect species. Any non-target insects that are vulnerable to Bt toxin will be affected if they eat any part of the GM crop. There was considerable media coverage when Monarch butterfly caterpillars that were fed only on Bt maize pollen in a laboratory experiment died (See Box 4). Another species that may be affected directly by Bt crops is the honey bee (*Apis mellifera*). At high doses Bt is toxic to bees but pollen from GM plants is unlikely to reach the doses required. Research with most of the widely grown commercial crops has found no effect on colony performance.

Studies in general did not show the effect of Bt-endotoxin on non-target organisms. Bt protein is relatively unstable so it will not remain or build-up in the food chain. Commonly, predators and parasites reared on insects feeding on GM plants do not grow to the same weight as those reared on insects feeding on non-GM plants. This is probably because there are lesser insects available on Bt-defended plants or because Bt-exposed insects are nutritionally poor for the predator/parasite.

One study, with lacewing, found the larvae died when they were fed prey raised on Bt maize. The researchers suggested Bt, which is normally not toxic to lacewings, had become toxic during processing by the lacewings' prey. It seems much more likely the prey used (*Spodoptera littoralis*) was not a good food for lacewing. Not all interactions will result in negative impacts. Bt maize, for example has reduced levels of insect damaged tissue, therefore it is less infestated by fungi that produce the mycotoxins that are toxic to humans and domestic animals.

Yet another transgenic plant has been shown to harm beneficial insects up the food-chain. Ladybirds fed on aphids that have eaten transgenic potato with snow-drop lectin lived half as long, laid 38% fewer eggs that were 4 times more likely to be unfertilized and 3 times less likely to hatch. This transgenic potato has now been revealed to be highly toxic to rats, and is most probably harmful to small mammals in the wild (Birch et al., 1997)

Box 4

Bt pollen effect on monarch butterfly

Losey et al. (1999) reported that monarch butterfly (Danaus plexippus) larvae reared on leaves of milkweed that were dusted with Bt maize pollen grew slowly and mortality rate was high as compared to larvae fed on normal maize pollen. This study raised concerns on conservation of monarch butterflies in USA and was widely and incorrectly interpreted to mean that GM crops were threatening non-pest insects. But this report was contradicted when follow up studies showed the effect of GM pollen on non-target insects, including the monarch butterfly to be negligible under real life conditions. In the first place, corn pollen does not fly long distances because it is pretty heavy. Also, most monarchs are moving at different times of the season when there's no corn pollen. It is argued that some of them might get killed by Bt corn pollen, but how many get killed when they are sprayed with insecticides. Further the scientific community discounted the original report because 1) it was conducted under artificial lab conditions, 2) the larvae were allowed to eat only corn pollen (which they don't often encounter in the open environment) and 3) there was no comparison group of larvae fed on ordinary corn pollen sprayed with regular Bt insecticide. The question is how much Bt corn pollen does it take to cause toxic effects in monarch caterpillars? It has been reported that monarch larvae eating leaves with pollen coating densities below 1000 grains/cm^2 had no effect on caterpillars weight or survival rate. Another question is what are the chances for caterpillars to encounter that dose under natural conditions? The answer is on an average less than 30% of the pollen that corn produces ends up on milkweed leaves. Data pooled from various locations showed that average density of Bt corn pollen density was about 170 grains/cm^2 and it rarely went above 600 grains/cm^2. Thus, given the low toxicity of Bt corn pollen and the low rates of exposure, the effect of Bt corn pollen from common commercial Bt hybrids on monarch butterflies is negligible.

10.5 Persistence of the Transgene or Transgene Product

The gene transferred into an organism or the resultant product can actually remain in environment leading to environmental problems. For example, in case of Bt crops it was suspected that insecticidal proteins can persist in the environment but experiments have proved that these are degraded in the soil. GM herbicide resistant oil seed rape, maize, sugarbeet and GM potato expressing either Bt toxin or pea lectin as mentioned earlier also did not survive well outside the agricultural field and did not take a weedy character in UK where field trials were conducted over a 10 year period. There are also concerns in case of microorganisms about their capacity to adapt to new environment conditions and persist in the environment as spores. It has been suggested that transgenic volunteers may persist in the field and become weeds due to their increased fitness. However there are no reports from many field trials and large scale cultivation of transgenic crop varieties that a transgenic has become a weed.

10.6 GM Crops Affect the Purity of Other Crops

Conventional non-GM crops will inevitably receive transgenes from GM crops, resulting in situations that are either undesired or unlawful. This has already happened in the case of GM Starlink maize containing the cry9C gene which was found in non-GM maize grains in the US (See Box 1). The organic farming industry is also particularly concerned about genetic mixing through pollen dispersal and mixing of seed. Liability may become a major issue.

10.7 Increased Use of Chemicals

Most of the transgenic varieties released for cultivation are against a particular herbicide and the farmers have to use a particular herbicide e.g. Roundup (Trade name of Monsanto herbicide) for a transgenic variety Roundup Ready Soybean. It is reported that farmers growing such varieties used a specific Roundup herbicide 2-5 times more in per unit area of land as compared to other weed management programmes. This increased use of herbicide is due to the fact that there is increased tolerance to Roundup of key weed species.

10.8 Effect on Rhizosphere and Microflora

Transgenic plants can influence the composition of microflora in their root zones or rhizospheres. Field studies with *B. napus* transgenic expressing barnase and barstar genes and non transgenic lines showed comparable rhizospheres qualitatively and quantitatively. Similarly, there was no difference in the bacterial cell consortia extracted from rhizosphere of non-transgenic and transgenic expressing cry1A(b) maize plants (Baumgrarte et al. 2004). But Andreote et al. (2004) observed significant differences in the composition of bacterial communities associated with the rhizosphere of non-transgenic and transgenic (expressing genes cab and nptII) plants of tobacco and eucalyptus.

10.9 Unpredictable Gene Expression or Transgene Instability

There is considerable evidence that transgenes show instability because they become deleted or modified following their integration in to the plant genome. The level of transgene expression may also decline with the passage of generations as gene silencing may occur. If and when these phenomena occur the crop becomes vulnerable to such stresses for which the transgenes provided protection. Unintended genomic changes can also occur as a secondary consequence of genetic modification. Such changes can lead to production of new proteins that may be toxic or allergenic or may disrupt or alter metabolic pathway that play a role in making the GMO successful.

Box 5

**Multinational companies disregard the rights of farmers e.g.,
Percy Schmeiser vs. Monsanto)**

Percy Schmeiser became the focus of international attention after he was taken to court by Monsanto, a biotech seed company, for using their patented Roundup Ready Canola seeds illegally. Schmeiser claimed pollen had drifted onto his property and he had merely replanted its seed next season. The Canadian court studied the report of samples from the 1,030 acres of canola planted and grown by Schmeiser Enterprises, Ltd. It consisted of 95% to 98% pure Roundup Ready plants as determined by independent testing, which could not have come from natural causes such as wind drift, spillage, etc. There were no close neighbors who grew Roundup Ready Canola when Schmeiser's lawyers claimed wind blew it on the enterprises, Ltd. Farms. The closest neighbour was five miles away. The judge explicitly rejected the pollen-drift theory because the facts did not support it. The court ruled that Schmeiser knew or should have known he was illegally using the protected plant variety and ordered him to pay damages of $40,000. Percy Schmeiser and his litigation with Monsanto had become familiar to the Commission long before his appearance as a witness for the Bio Dynamic Farming and Gardening Association. Anti GMO campaigners mentioned his case as an example of the perceived evils of genetic modification business.

11. International Protocols and Conventions on Biosafety

11.1 Cartagena Protocol on Biosafety

There have been various conventions, protocols and treaties for safe use of GMOs and their products. A Convention on Biological Diversity (CBD) was adopted in June 1992 which came into force in 1993. The objectives of CBD are "the conservation of biological diversity, the sustainable use of its components and access and benefit sharing arising out of utilization of biological resources. Under this convention contracting parties agreed to consider and develop appropriate procedures to address the safe transfer, handling and use of any living modified organisms (LMOs). CBD recognized that biotechnology inventions may have adverse effects on conservation and sustainable use of biological diversity (Article 19.3 of CBD) and a biosafety protocol is the result of that process. A biosafety protocol named as Cartagena protocol on biosafety to the Convention on Biological Diversity was finalized in February 1999 at

Cartagena, Colombia but adopted a year later on 29 January, 2000 in Montreal Canada. Protocol entered into force in September 11, 2003. 147 countries have ratified the protocol as of May 2008. India signed the Cartagena Protocol on Biosafety on 23rd January 2003, which aims at ensuring an adequate level of protection in transfer, handling and use of genetically improved organisms, particularly during their trans-boundary movement. The protocol features two separate set of procedures one for LMOs that are to be intentionally introduced into the environment and one for those that are to be used directly as food or feed or for processing. It incorporates the use of precautionary principle, the application of the advance informed agreement (AIA) procedure for import of the organisms, risk assessment, risk management cooperation in preventive and emergency measures, capacity building and exchange of scientific, technical, environmental and legal information regarding the organisms through a biosafety clearing house mechanism. Various elements of the protocol which needed attention are as follows:

i. **Advance informed agreement procedure**: Under the biosafety protocol, the most rigorous procedures are reserved for GMOs that are to be introduced intentionally into the environment. These include seeds, live fish and other organisms that are destined to grow and that have the potential to pass their modified genes on to succeeding generations. It includes four components: notification by the Party of export or the exporter, acknowledgment of receipt of notification by the Party of import, decision procedure and review of decisions. The purpose of this procedure is to ensure that importing countries have both the opportunity and the capacity to assess risks that may be associated with the LMO before agreeing to its import.

Specifically, the exporter starts by providing the government of importing country a detailed, written description of the LMO in advance of the first shipment. A Competent National Authority in the importing country is to acknowledge receipt of notification. The decision could be (i) approving the import, (ii) prohibiting the import, (iii) requesting additional relevant information, or (iv)extending the 270 days by a defined period of time.

Except in a case in which consent is unconditional, in other cases the Party of import must indicate the reasons on which its decision are based. The absence of response is not to be interpreted as implying consent.

A party of import may, at any time, in light of new scientific information, review its decisions. However, the Protocol's AIA procedure does not apply to certain categories of LMOs:i.e.

- LMOs in transit;
- LMOs destined for contained use;
- LMOs intended for direct use as food or feed or for processing

It should be noted that, while the Protocol's AIA procedure does not apply to certain categories of LMOs, Parties have the right to regulate the importation on the basis of domestic legislation. In this way the AIA Procedure ensures that recipient countries have the opportunity to assess any risks that may be associated with LMO before agreeing to its import.

ii. **Procedures for LMOs intended for direct use as food or feed for processing (LMOS-FFP)**: LMOs intended for direct use as food or feed, or processing (LMOs-FFP) and not as seeds for growing new crops represent a large category of agricultural commodities.

   Instead of requiring the use of the AIA procedure, the Protocol establishes a simpler system for the transboundary movement of LMOs-FFPs. Under this procedure, governments that approve these commodities for domestic use must communicate this decision to the world community via the Biosafety Clearing-House within 15 days of its decision. They must also provide detailed information about their decision.

   Decisions by an importing country on whether or not to import these LMO-FFPs are taken under its domestic regulatory framework. In the absence of domestic regulatory framework a country may declare through the biosafety Clearing House that its decisions on the first import of LMOs-FFP will be taken in accordance with risk assessment as set out in the protocol and timeframe for decisions making. In case of insufficient relevant scientific information and knowledge, the importing country may use precaution in making their decisions on the import of LMOs-FFP.

iii. **Risk assessment**: The protocol empowers governments to make decisions in accordance with scientifically sound risk assessments. These assessments aim to identify and evaluate the potential adverse effects that a LMO may have on the conservation and sustainable use of biodiversity in the receiving environments. They are to be undertaken in a scientific manner using recognized risk assessment techniques. A country considering permitting the import of a GMO is responsible for ensuring that a risk assessment is carried out, but it also has the right to require the exporter to do the work or bear the cost.

iv. **Risk management and emergency procedures**: The Protocol requires each country to manage and control any risks that may be identified by a risk assessment. Key elements of risk management include monitoring systems, research programs, technical training and improved domestic coordination amongst government agencies and services. The protocol also requires each government to notify and consult other affected governments when it becomes aware that LMOs under its jurisdiction may cross international borders due to illegal trade or release into the environment.

This will enable them to pursue emergency measures or other appropriate action. Governments must enable them to pursue emergency measures or other appropriate action. Governments must establish official contact points for emergencies as a way of improving international coordination.

v. **Export documentation**: For GMOs intended for direct introduction into the environment, the accompanying documentation must clearly state that the shipment contains GMOs. It must specify the identity and relevant traits and characteristics of the GMO; any requirements for its safe handling, storage, transport and use; a contact point for further information and the names and addresses of the importer and exporter. In cases where a government agrees to import a genetically modified commodity intended for direct use as food or feed or for processing, the shipment must clearly indicate that it "may contain" living modified organisms and that these organisms are not intended for introduction into the environment.

vi. **The Biosafety Clearing House (BCH)**: The Biosafety Clearing House contains information on national laws, regulation, and guidelines for implementing the Protocol. The Biosafety Clearing-House also includes information required under the AIA procedure, summaries of risk assessments and environmental reviews, bilateral and multilateral agreements, reports on efforts to implement the Protocol, plus other scientific, legal, environmental and technical information. The Biosafety Clearing-House and has been developed largely as an Internet-based system and can be found at http://bch.biodiv.org.

vii. **Unintentional transboundary movement of LMOs**: When a country knows of an unintentional transboundary movement of LMOs that is likely o have significant adverse effects on biodiversity and human health, it must notify affected or potentially affected States, the Biosafety Clearly House and relevant international organizations regarding information on the unintentional release. Countries must initiate immediate consultation with the affected or potentially affected States to enable them to determine response an emergency measures.

viii. **Capacity-building and finance**: Parties are encouraged to assist with scientific and technical training and to promote the transfer of technology, know-how, and financial resources.

ix. **Public awareness and participation**: It is clearly important that individual citizens understand and are involved in national decisions on GMOs. The Protocol therefore calls for cooperation on promoting public awareness of the safe transfer, handling and use of GMOs. It specifically highlights the need for education, which will increasingly have to address GMOs as biotechnology becomes more and more a part of our lives. The Protocol also calls for the public to be actively consulted on GMOs and biosafety. Individuals, communities and nongovernmental organizations should remain fully engaged in this complex issue. This will enable people to contribute to the final decisions taken by governments, thus promoting transparency and informed decision-making.

x. **Issue of non-Parties**: The Protocol addresses the obligations of Parties in relation to the trans-boundary movements of LMOs to and from non-Parties to the Protocol. The trans-boundary movements between Parties and non-Parties must be carried out in a manner that is consistent with the objective of the Protocol. Parties are required to encourage non-Parties to adhere to the Protocol and to contribute information to the Biosafety Clearing-House. (Article 24).

xi. **Institutional arrangement at the national level**: Parties are required to designate national institutions to perform functions relating to the Protocol. Each party needs to designate one national focal pint to be responsible on its behalf for liaison with the Secretariat. Each Party also needs to designate one or more competent national authorities, which are responsible for performing the administrative functions required by the Protocol and which shall be authorized to act on its behalf with respect to those functions. A Party may designate a single entity to fulfill the functions of both focal point and competent national authority.

11.2 WTO and Other International Agreements

Cartagena protocol is the only international agreement which deals exclusively with GMOs, but there are number of other agreements, which address various aspects of biosafety.

i. International Plant Protection Convention (IPPC) protects plant health by assessing and managing the risks of plant pests. The IPPC is in the process of setting standards to address the plant pest risks associated with GMOs and invade species. Any GMO that could be considered as a plant pest falls within the scope of this Treaty. The IPPC allows governments to take action to prevent the introduction and spread of such pests. It established procedures for analyzing pest risks, including impacts on natural vegetation.

ii. Codex Alimentarius Commission addresses food safety and consumer health. On March 8, 2002 at Yokohama, Japan a task force of the Codex Alimentarius Commission has reached an agreement on a final draft of "Principles for the risk analysis of foods derived from Biotechnology". The Commission is also considering the issue of labeling biotech foods to allow the consumer to make an informed choice. The Codex Alimentarius, or the food code, has become the seminal global reference point for consumers, food producers and processors, national food control agencies and the international food trade.

iii. World Organization for Animal Health referred as Office of the International des Epizootics, (OIE), which develops standards and guidelines designed to prevent the introduction of infectious agents and diseases into the importing country during international trade in animals, animal genetic material and animal products.

iv. World Trade Organization (WTO) stipulates a number of WTO agreements, such as the Agreement on the Application of Sanitary and Phytosanitary Measures (SPS) and the Technical Barriers to Trade (TBT) Agreement contain provisions that are relevant to biosafety.

Whereas both the Cartagena Protocol on Biosafety and WTO advocate the use of science based risk assessments as a means to justify trade related measures, there are areas of potential conflicts mainly with respect to application of precautionary approach in the conditions that scientific evidence is insufficient and the socio-economic considerations in the decision of importing LMOs.

For example with regard to a certain biotechnology product, the details needed to be considered for risk assessment under the protocol and the SPS are very different. The SPS does not specify exactly what a risk assessment is, but the protocol elaborates this in detail. Further, the SPS does not mention risk management, but merely risk assessment. The scope of socio-economic considerations under the Protocol is wide, while SPS puts strict limits on the economic considerations. The mandatory labeling of LMOs-FFPs under the Protocol may also be in conflict with the WTO rules.

12. National Biosafety Regulatory Framework in India

Govt. of India has evolved a comprehensive regulatory mechanism for development and evaluation of GMOs and rDNA research work. The Ministry of Environment and Forests (MoEF) is the nodal agency for release of GMOs in the country. The Ministry has enacted Environment and Protection Act (EPA), 1986, rules 1989, to provide for protection and improvement of environment and the related matters. The rules and regulations cover the areas of research as well as large scale applications of GMOs and products made there from. The rules also cover the application of hazardous microorganisms which may not be genetically modified. Department of Biotechnology (DBT) had formulated recombinant DNA Guidelines in 1990 which were revised in 1994. There are six competent authorities for the regulatory mechanism as described below and shown in Fig. 14.1. The salient features are.

12.1 Recombinant DNA Advisory Committee (RDAC)

A committee constituted by DBT referred as RDAC, take note of developments in biotechnology at national and international levels and recommends safety regulations for research and applications.

12.2 Institutional Biosafety Committee (IBSC)

It is the nodal point for interaction within the institution for implementation of guidelines. For this, institution carrying out research activities on genetic manipulation should constitute IBSC with one DBT nominee. The main

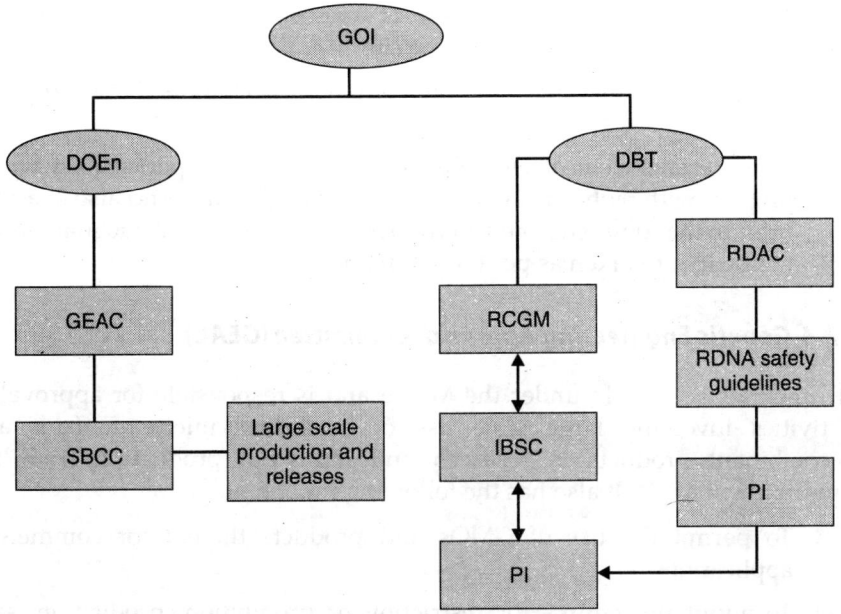

Fig. 14.1: GOI-Government of India; DBT-Department of Biotechnology; RDAC- Recombinant DNA Advisory Committee; IBSC-Institutional Biosafety Committee; RCGM-Review Committee on Genetic Manipulation; DOEn-Department of Environment; GEAC-Genetic Engineering Appraisal Committee; SBCC-State Biotechnology Coordination Committee; PI-Principal Investigator (R&D/Industry/Others); FA-Funding Agency (Govt./Private & Public Institutions)

activities are: i) to note and to approve r-DNA work; ii) to ensure adherence of r-DNA safety guidelines of government; iii) to prepare emergency plan according to guidelines; iv) to recommend to RCGM about category III risk or above experiments and to seek RCGM's approval; v) to act as nodal point for interaction with statutory bodies; vi) to ensure experimentation at designated locations, taking into account approved protocols.

12.3 Review Committee on Genetic Manipulation (RCGM)

DBT has next higher level of body known as RCGM which has the following functions:

i. To bring out manuals of guidelines specifying procedures for regulatory process on GMOs in research, use and applications including industry with a view to ensure environmental safety.

ii. To review all the work going on r-DNA projects involving high risk category and controlled field experiments.

iii. To lay down procedures for restriction or prohibition, production, sale, import and use of GMOs both for research and applications.

iv. To permit experiments with category III risks and above with appropriate containment.

    v. To authorize field experiments in 20 acres in multi-locations in one crop season with up to one acre at one site.

    vi. To generate relevant data on transgenic materials in appropriate systems.

    vii. To undertake visits of sites of experimental facilities periodically where projects with biohazard potentials are being pursued and also at a time prior to the commencement of the activity to ensure that adequate safety measures are taken as per the guidelines.

12.4 Genetic Engineering Appraisal Committee (GEAC)

It functions as a body under the MOEF and is responsible for approval of activities involving large scale use of hazardous microorganisms and recombinant products in research and industrial production from the environment angle. It also has the following functions:

    i. To permit the use of GMOs and products thereof for commercial applications.

    ii. To adopt procedures for restriction or prohibition, production, sale, import and use of GMOs both for research and applications under EPA, 1986.

    iii. To authorize large scale production and release of GMOs and products thereof into the environment.

    iv. To authorize agencies or persons to have powers to take punitive actions under the EPA.

12.5 State Biotechnology Coordination Committee (SBCC)

In each state there is a State Biotechnology Coordination committee (SBCC) headed by the Chief Secretary where research and applications of GMOs are contemplated.

12.6 District Level Committee (DLC)

DLC is the district level committee headed by district collector as an authoritative unit to monitor safety regulations. Both SBCC and DLC work along with RCGM in the inspection and monitoring of the experiments at the field sites.

13. Recombinant DNA Safety Guidelines

DBT had formulated recombinant DNA Guidelines in 1990 which were revised in 1994. It include guidelines for R&D activities on GMOs, transgenic crops, large-scale production and deliberate release of GMOs, plants, animals and products into the environment, shipment and importation of GMOs for laboratory research. The issues relating to genetic engineering of human embryo, use of embryos and fetuses research and human germ line, and gene therapy areas have not been considered while framing the guidelines.

Four different biosafety levels have been recognized and containment facilities for each level are recommended which have already been explained. Recombinant DNA/ genetic engineering research activities guidelines have been grouped under three categories:

Category I: It includes: i) Routine cloning of defined genes, defined non-coding stretches of DNA and open reading frames in defined genes in *E. coli* or other bacterial/fungal hosts which are generally considered as safe to human, animals and plants. ii) Transfer of defined cloned genes into *Agrobacterium*. iii) Use of defined reporter genes to study transient expression in plant cells and to study genetic transformation conditions. iv) Molecular analysis of transgenic plants grown *in vitro*. It also includes those experiments which involved self cloning using strains and also interspecies cloning belonging to organisms in the same group. The experiments need intimation to IBSC in the prescribed proforma.

Category II: It include experiments carried out in the Lab and green house/net house using defined DNA fragments non-pathogenic to human and animals for genetic transformation of plants, both model species and crop species and the plants are grown in green house/net house for molecular and phenotypic evaluation. Permission to perform experiments will be provided by IBSC. It would be intimated to RCGM before execution of experiments.

Category III: It includes experiments having high risk where the escape of transgenic traits into the open environment could cause significant alterations in the biosphere, the ecosystem, plants and animals by dispersing new genetic traits the effects of which cannot be judged precisely. Further this also includes experiments conducted in green house and open field conditions not belonging to the above category II types. Such experiments could be conducted only after the clearance from RCGM and notified by DBT. It also includes experiments on toxin gene cloning, cloning of genes for vaccine production, etc.

The controlled release of GMOs should be done under appropriate containment facilities to ensure safety and to prevent unwanted release in the environment

Pre-release tests of GMOs in agriculture should include elucidation of requirements for vegetative growth and persistence and stability in small plots and experimental fields.

14. Revised Guidelines for Research in Transgenic Plants, 1998

DBT brought out separate guidelines for research in transgenic plants in 1998 which include the guidelines for toxicity and allergenicity of transgenic seeds, plants and plant parts. The genetic engineering experiments on plants have been grouped under three categories. Category I includes routine cloning of defined genes, defined non-coding stretches of DNA and open reading frames in defined genes in *E. coli* or other bacterial/fungal hosts which are generally considered as safe to human, animals and plants. The category II experiments include experiments carried out in lab and greenhouse/net house using defined DNA fragments non-pathogenic to human and animals for genetic transformation of plants, both model species and crop species. Category III includes experiments having high risk where the escape of transgenic traits into the open environment could cause significant alterations in the biosphere, the ecosystem, plants and animals by dispersing new genetic traits the effects of which cannot be judged precisely. Further this also includes experiments conducted in green house and open field conditions having risks mentioned above.

Risk management is employed during the development and evaluation of an organism in a systematic fashion in the laboratory, through stages of field testing to commercialization. Right from the initiation of research, risk assessment and management considerations should be kept in mind. Planned field experiments with transgenic plants are permitted only after a step-wise (laboratory to growth chamber and greenhouse) evaluation, either in India or elsewhere, to generate data on the following: i) Characteristics of the donor organisms providing the target nucleic acids; ii) Characteristics of the vectors used; iii) Characteristics of the transgenic inserts which includes specific functions coded by the inserted nucleic stretches including the marker gene inserts, expression of nucleic acid products and their activities, toxicity and allergenicity of nucleic acid products to human and animals sequence, iv) regulatory mechanism utilized in the expression of cassette, v) cell lines used for shuttling and amplification of the cassette vi) Characteristics of the transgenic plants which include the methods for detection of transgenic plants, methods of detection and characterization of the escaped transgenic traits in the environment, toxicity and pathogenicity of transgenic plants and their fruits to other plants in the ecosystem and environment, possibility of and the extent of transgenic escape and pollen transfer to wild near relatives and pathogenicity, toxicity and allergenicity of the transgenic plants and their fruits to human and animals. In addition, the following is also insisted upon.

(a) Laboratory data to show that protein products of transgenes are safe to the environment and human beings.

(b) Isolation distance as applicable to foundation seed of the crop be provided to transgenic crops when grown in the field.

(c) A few rows of the same crop as the transgenic one should be planted beyond the isolation distance to act as pollen trap.

(d) Non-transgenic plants should be grown within the isolation distance at 1 or 5 m intervals to determine the distance of pollen escape.

(e) All the vegetative plants and left-over seeds must be destroyed by burning after the conduct of experiments.

(f) After the experiment, the land may be left fallow and all plants that emerge must be destroyed.

(g) The experimental field may be visited by the company authorized personnel only, and all records of visits are to be maintained.

(h) Full account of transgenic seeds produced is to be kept and no part of this seed lot can be transacted or further propagated without authorization as per the guidelines.

To monitor over a period of time, the impact of transgenic plants on the environment, a special Monitoring cum Evaluation Committee (MEC) has been set up by the RCGM. The committee undertakes field visits at the experimental sites and suggests remedial measures to adjust the trial design, if required, based on the on-spot situation. This committee also collects and reviews the information on the comparative agronomic advantages of the transgenic plants and advises the RCGM on the risks and benefits from the use of transgenic plants and advises the RCGM on the risks and benefits from the use of transgenic plants put into evaluation. Trials will be done for at least one year with minimum of four replications and ten locations in the agroecological zone. The biological advantage of transgenic will be judged and communicated by RCGM to GEAC for consideration of release in to the environment.

The guidelines include complete design of a contained green house suitable for conducting research with transgenic plants. Besides, it provides the basis for generating food safety information on transgenic plants and plant parts.

A separate section (No. 6) in the Seed Policy, 2002 on transgenic plant varieties has been put in place. Under this all genetically engineered crops/ varieties will be tested for environment and biosafety before their commercial release. Seeds of transgenic plant varieties for research purposes will be imported only through National Bureau of Plant Genetic Resources as per EPA Act, 1986. Clearance for import of transgenic material for research purposes would be provided by RCGM. Transgenic varieties will be tested for 2 seasons for their agronomic values under All India coordinated project trials of ICAR, in coordination with the tests for environment and biosafety clearance as per EPA Act before any variety is commercially released. Transgenic plant varieties cannot be protected by patent regime in India but can be registered under Protection of Plant Varieties and Farmers' Rights (PPV&FR) Act, 2001 in the same manner as non-transgenic plant variety.

Besides, Drug and Cosmetic rules (8th amendment), 1988 and Drug Policy, 2002 for all recombinant products considered to be new products, and guidelines for generating preclinical and clinical data for rDNA therapeutics, 1999 are under operation.

15. Cross Border Movement of Transgenic Germplasm for Research Purposes

As per the revised guidelines for research in transgenic plants, 1998, clearance for import of transgenic material, for research purposes would be provided by the RCGM to the concerned importer applicant. The RCGM will issue an import certificate after looking into the documents related to the safety of the material and the national need. The RCGM will take into consideration the facilities available with the importer for in-soil tests on the transgenic material.

The importer of a transgenic material may import the material accompanied by an appropriate phyto-sanitary certificate issued by the authority of the country of export. However, such import may be routed through the Director, NBPGR on the basis of the import permit issued by the RCGM of DBT. The import certificate would be cancelled if NBPGR is not provided the phyto-sanitary certificate. NBPGR will provide information on the time that is required for phyto-sanitary evaluation. These evaluations will be done in a time- bound manner in presence of the agents of the institutes or the commercial organizations that are importing the material, if they so desire. Parts of the seed material will be kept at NBPGR in double lock system in the presence of the importer. This lot of seed will act as a source material in case of any legal dispute.

16. Regulatory Frameworks in Different Countries

16.1 USA

The USDA's Animal and Plant Health Inspection Services (APHIS) is the lead agency for the regulation of genetically engineered plants including the experimental evaluation of these products in confined field trials. The Environmental Protection Agency (EPA) is responsible for assuring the human and environmental safety of pesticidal substances engineered into plants, and the Food and Drug Administration (FDA) is responsible for assuring that foods and drugs derived from genetic engineered are as safe as their traditional counterparts. Products are generally regulated according to their intended use, with some products being regulated under more than one agency e.g. pesticidal plants. All the three regulatory agencies have the legal power to demand immediate removal from the marketplace of any product post commercialization if any new and valid data indicates safety concerns on consumers or the environment. Regarding research and development activities, compliance with National Institute of Health guidelines is mandatory for working with GMOs for all scientists receiving federal funding or working for federal agencies.

16.2 European Union

The deliberate release of GMOs into the environment is under the Directive 2001/18/EC. The Directive puts in place a step by step approval process on a case by case assessment of the risks to human health and the environment before any GMO or product consisting of, or containing GMOs can be released to the environment or placed on the market. The approval process requires submission of the notification to the competent authority in a member state of European Union (EU) where the GMO will be field tested/marketed. The member state gives a summary/assessment report for the EU commission and competent authorities of all member states. Besides, public is also provided an opportunity to provide comments, which are discussed in an attempt to reach agreement. At the end of review process, the competent authority provides written consent for marketing the GMO for a period of no more than 10 years. The period of validity, the conditions for marketing the product, the labeling and monitoring requirements are all specified in the consent. The Directive requires that the labeling should include the words 'this product contains genetically modified organisms'.

16.3 Canada

Canadian Food Inspection Agency (CFIA), Health Canada and Environment Canada co-ordinates the regulations of the biotechnology products. The CFIA is responsible for regulating the import, environmental release, variety registration and use in livestock feeds of plants with novel traits. Health Canada is solely responsible for assessing the human health safety of foods. Environment Canada is responsible for administering the new substances notification regulations and for performing environmental risk assessment of toxic substances, including organisms and microorganisms that may have been derived from biotechnology.

Canada is the only country where regulatory system is operative on the novelty of traits expressed by plants or the novel attributes of foods or food ingredients, irrespective of the means by which the novel traits were introduced. The Canadian regulatory system refers to plants with novel traits (PNT) and novel foods in place of GM plants or GM foods. Under this regime, all agricultural commodities and food products, whether they are produced using conventional technologies or biotechnologies, are governed under the same Acts. Depending on the type of product, the relevant piece of legislation is applied i.e. Seeds Act, Feeds Act, Fertilizers Act, Food and Drugs Act, Health of Animals Act, or the Canadian Environmental Protection Act (CEPA). For example, a herbicide tolerant Canola produced by genetic engineering or mutagenesis (established plant breeding tool) are subject to same environmental or food safety risk assessment although the latter approach has been in use for more than 80 years.

16.4 Australia

In Australia, research, manufacture, production, commercial release and import of GMOs are regulated under the Gene Technology Act, 2000 by the Gene Technology Regulator (GTR).

16.5 Argentina

Regulations concerning the environmental release of GMOs were developed by Commission Nacional Asesora de Biotecnologia Agropecuaria (The

National Advisory Committee on Agricultural Biosafety or CONABIA) and are enforced by Secretary of Agriculture, Livestock, Fisheries and Food (SAGPyA). The regulatory requirements for GMOs are based on guidelines in the form of non-legislative resolution that are integrated in the overall regulatory system and there is no specific law that makes the resolutions legally binding.

16.6 Asian Region

China's biosafety guidelines were produced by the State Science and Technology Commission in December 1993, under which the administrative responsibility for biosafety of various products has been assigned to the relevant administrative departments. In 2002 China has established rules on GMOs to strengthen the safety and management of GMO products. Besides other detailed procedures, these rules require all GM products to be labeled. Japan uses voluntary guidelines administered through four governmental agencies viz. Ministry of Science & Technology for lab work, Ministry of International Trade and Industry for industrial applications, Ministry of Agriculture, Forestry and Fisheries for safety of animal feeds, feed additives and environmental release of GMOs and Department of Health and Welfare for food and food additives produced by rDNA technology. Philippines and Malaysia have completed their biosafety guidelines. Thailand has already approved field testing of GMOs after finalization of regulations in 1993.

17. Socio-Economic Considerations

One of the most important considerations in commercialization of biotechnology based products is its socio-economic implications. Quite often, scientists might come up with techniques which *per se* may look novel and innovative, but from a global perspective may have wide-ranging socio-political and socio-economic ramifications. The development of Genetic Use Restriction Technologies (GURT) in the development of transgenic varieties is a prominent example. Though it looks appropriate for private sector but for the farmers of developing countries who rely of farm-saved seed is totally a bad thing and will be disastrous for such countries. India has totally banned the use of GURT in plant varieties for registration under Protection of Plant Varieties and Farmers' Rights Act, 2001. India is a land of small farm holdings. There are about 90 million operational farm holdings in the country, and about 60 percent of them own less than one hectare of land. Thirty five percent own land between one hectare to 4 hectare and only 5% own more than 4 hectares of land. Farming provides a livelihood to nearly 60% of the Indian population. Unless developing countries have policies in place to ensure that small farmers have access to better agricultural input delivery system, extension services, markets, and infra-structure. There will be increased inequality of income and wealth for small farmers if transgenics are introduced as big farmers are likely to capture most of the benefits through early adoption of the technology, expanded production, and reduced unit costs. However, in India farmers have adopted the biotech cotton technology and got benefits. Companies were selling the biotech seed at a much higher rate and the case was filed and hence Indian legal system intervened and the companies reduced the price. If benefits are to flow then regulatory system and legal system is also to be put in proper framework.

Another concern of the developing countries is the open liberalized and globalized markets where competition on plain-leveled field would often determine who is going to prevail or perish. Obviously, the one with efficient, effective and relevant technological intervention would be able to compete in terms of cost and quality of the product. Therefore, it would be the cutting edge of science and technology which decides as to who is going to attain and sustain advantages on short-term and long-term basis in the international markets. Some developing countries may be in a position to share significantly the benefits accruing from biotechnological developments on a long run. For instance, countries such as India, China, Mexico, Brazil, Thailand and Philippines share numerous important attributes including large domestic markets, a strong agricultural production and trading capacity, reasonably strong agricultural research network, a broad human capital base, and good technological and scientific capability in crop biotechnology. But, many other countries in the Asia and Africa are in a potential disadvantage position in effective exploitation of frontier technologies, and consequently, international trade, unless there are significant policy changes.

It is well known that MNCs have a dominant role in biotechnological research and the key technologies are in the 'hands of a few' peoples. Further, there is consolidation of seed industry where large MNCs have either purchased or are in the process of purchasing smaller seed and biotechnology companies. Consequently, the fear is that transgenic crops will prove to be expensive for resource-poor farmers. Considering the fact that crop biotechnology has a heavy involvement of private sector, particularly a few MNCs, these acquisitions and mergers have raised apprehensions and valid concerns that market considerations and mergers will greatly influence the areas as well as commodities chosen for research in biotechnology, besides strengthening monopolistic tendencies.

Public acceptance is one of the major hurdles for the adoption of the first wave of products of agricultural biotechnology. In many countries, people are naturally cautious about the transgenic cultivars and their products. In general, people around the world appear to accept biotechnology in medical applications more easily than biotechnology in the field of agriculture or food processing. In some countries, the public and the scientific community hold different views as in European Union, commercialization of genetically modified crops have faced stiff resistance whereas public acceptance is much better in the USA. There are several factors that can play a key role in public acceptance of genetically modified crops. Scientific demonstrations of biosafety of transgenic crops and reviews by government agencies are extremely important in gaining public acceptance. What role credible experts will play in communicating the issues to the public in a realistic and effective manner can make a huge difference. Public acceptance is also greatly determined by the kind of information provided by the media to the general public and various organizations concerned about farmers. Misinformed public debates on key issues related to crop biotechnology can result in erosion of public confidence and can create mistrust in the technology and its developers, irrespective whether the developers are from the public or private sector. Clear and understandable consumer information is a very important part of the public acceptance process. Besides media, research organizations and scientific institution concerned with crop improvement must also take up the responsibility in bringing awareness in public about the applications of genetic engineering in agriculture, their potential benefits as well as constraints (Rai and Prasanna, 2000)

Selected reading

Anonymous 2004. National Consultations on Biosafety aspects related to Genetically Modified Organisms, BCIL, Delhi and DBT, India, pp. 338.

Chawla, H.S. (2009) Introduction to Plant Biotechnology (3rd edition). Science Publishers Inc., USA and Oxford & IBH Publishers, Delhi, India, pp 728

Franck-Oberaspach, S.L. and Keller, B. 1997. Consequence of classical and biotechnological resistance breeding for food toxicology and allergenicity. *Plant Breed.* 116: 1-17

Malik, V.S. and Saroha, M.K. 1999. Marker gene controversy in transgenic plants. *J. Plant Biochem. Biotechnol* 8: 1-13.

McAllan, A.B. and Smith, R.H. 1973. Degradation of nucleic acids in the rumen. *British J. Nutrition* 29: 467-474.

Messeguer, J. 2003. Gene flow assessment in transgenic plants. *Plant Cell Tissue Organ Cult.* 73: 201-212.

Nap, J.P., Bijvoet, J. and Strikema, W.J. 1992. Biosafety of kanamycin resistant transgenic plants: an overview. *Transgenic Crop* 1:239-249

Rai, M. and Prasanna, B.M. 2000. Transgenics in Agriculture. Indian Council of Agricultural Research, Delhi, pp. 142.

Singh, B.D. 2006. Plant Biotechnology, Kalyani Publishers, India, pp. 755.

Bioethics

Ethics is the discipline concerned with what is good or bad, right or wrong. It has theoretical and practical aspects. Ethics seeks to establish norms or standards of conduct (normative ethics), and to analyze the basis of judgments about what is right and wrong (descriptive ethics). Applied or practical ethics is the application of theoretical ethical tools and ethical norms to address actual moral choices.

Bioethics deals with the ethical implications of biological research, biotechnological research and the biological and medical applications of research. Specific bioethics issues arise in debates over the dignity of the human being, beginning-of-life and end-of-life issues, consent to medical treatment, consent to use of germplasm, consent for use of GMO's, freedom of research, the consent of the donor of human genetic material, access to health care and distribution of health resources, and equitable access to the outcomes of biological research, as well as animal protection and environmental ethics.

1. Ethics v/s Morality

'Ethics' and 'morality' are often used interchangeably, but they do have different aspects. For instance, practical ethics aims to guide right behavior while 'morality' refers to the underlying moral values that are used to assess what is right and wrong. In the field of IP, some patent laws refer to inventions the exploitation of which would be contrary to *ordre public* or morality, and some trademark laws refer to trademarks that are contrary to morality. In this sense, 'morality' could refer to the shared values of a community, values that might differ from one community to another.

2. Law v/s Ethics

Law and ethics are closely interrelated, but they are not the same thing. Some of the acts that are legal might be considered unethical. As a simple example, it is normally unethical to tell a lie, but only in some circumstances is it a true crime. There can be strong commonality and consistency between the law of human rights, and ethical norms and expectations, but it would actually reduce the legal effect and status of human rights law to regard it as giving ethical guidance only. Sometimes legislators choose not to pass laws on certain issues, as a conscious choice to allow communities' ethical considerations to govern behaviour, instead of legal rules. Certain forms of stem cell research may not actually break the law of a particular country, but some might still argue that it is unethical.

3. The Ethical Basis of IP Policy

In principle, appropriate IP protection aims to promote policy objectives that are consistent with widely accepted ethical principles. But there are different ways of analyzing the ethical basis of IP laws. Some IP laws and principles are argued to have a 'natural rights' basis, reflecting an inherent entitlement to just reward and recognition for one's intellectual and creative contributions.

On the other hand, there is also a strong utilitarian flavor to IP law and policy, as a conscious tool to promote social welfare. A utilitarian approach to ethics would assess moral value of a measure or an action according to its contribution to overall social utility or welfare. This utilitarian ethic is increasingly emphasized in current debate on IP as a tool of public policy. It is echoed in the TRIPs Agreement, which provides that the protection and enforcement of IP rights should 'contribute to the promotion of technological innovation and to the transfer and dissemination of technology, to the mutual advantage of producers and users of technological knowledge and in a manner conducive to social and economic welfare, and to a balance of rights and obligations.'

4. General Principles

There are four broad principles in bioethical concerns, namely: transparency, prior informed consent, equitable benefit-sharing, and pluralism.

4.1 Transparency

Transparency and access to knowledge are key principles that are both central to bioethical concerns and facilitate the ethical scrutiny of new technologies. The Universal Declaration on Bioethics and Human Rights (UDBHR) calls for the greatest possible flow and the rapid sharing of knowledge concerning medical, scientific and technological developments. As a matter of principle, the patent system is required to promote the flow of timely information about new technologies. It is through the patent system that a new biotechnology is first, and most fully, disclosed to the public. Additionally, patent documents reveal the identity of inventors, commercial enterprises, as well as governmental and educational institutes that are involved in the creation and development of those technologies. Patent information systems shed light at an early stage on the development of technologies that may have important bioethical implications. They may be used to monitor:

- overall trends and patterns in the development of key technologies (for example, the trends in patenting gene sequences),
- state of the art and recent developments in a particular technology area (such as recent technologies concerning stem cells),
- research and patenting activities of specific firms, institutes and individuals.

The transparency of the patent system therefore supports ethical scrutiny of biotechnology and can help inform the bioethics debate. Policymakers and others concerned with bioethics issues may need further distillation and analysis of this raw patent data so that the broader implications can be assessed.

Access to information through the patent system does not, however, entail freedom to use that information in practice as technology – precisely because a patent gives exclusive rights over that technology in those countries where it has legal effect. How those exclusive rights are obtained and exercised can also have an ethical dimension?

4.2 Consent

Consent to use certain inputs to biotechnological research have been a recurrent issue with bioethical implications. There have been cases where genetic materials taken from the human body have been used as inputs for research, leading to inventions, which were subsequently patented. This has raised questions about the need to obtain the prior consent of the human subjects concerned, and whether consent extends to the patenting of outputs from research. For example the case of hairy cell leukemia. Dr. Golde patented a cell line established from Mr. Moore's discarded spleen tissues. A court was confronted with some difficult legal issues. Does consent to have medical treatment imply consent to use of cells in research? Does consent to allow research entail consent to patenting the results of that research? How do legal obligations, differing forms of property, and ethical expectations overlap? The Court, in Moore v/s Regents of the University of California, rejected Mr. Moore's claim to ownership interest in the patent, but ruled that a physician had a "fiduciary duty" to inform a patient of any economic or personal interest in using or studying his tissues.

Thus, consent is a key issue in bioethics, and it can be helpful to explore the relationship and the boundaries between legal and ethical aspects of consent to use genetic inputs to research. The issue of recognition of the interests of the donor of human genetic resources may overlap with questions of research involving human subjects in general. It may be necessary to clarify whether consent to take part in medical research or to undergo medical treatment extends to consent to the obtaining of IP based on that research. Various texts state the importance of the consent of the persons involved in research. The UDBHR (Article 6 (II)) provides that "scientific research should only be carried out with the prior, free, express and informed consent of the person concerned".

A similar debate applies to other genetic resources, such as genetic resources obtained through bioprospecting, which are subsequently used in research to create new technologies for which patent protection may be sought. The Convention on Biological Diversity (CBD) makes prior informed consent a condition of access to genetic material of plant, animal or microbial origin. While the UDBHR sets prior informed consent in the context of human dignity and autonomy, the CBD links it to the sovereignty of nations over their resources, and the interests of indigenous and local communities.

4.3 Equitable Sharing of Benefits

A further theme is how the benefits of research should be shared, and what it means for the sharing to be equitable. This theme potentially has both legal and ethical aspects. Human rights law – as expressed in the Universal Declaration on Human Rights – accords everyone the right "to share in scientific advancement and its benefits" while affirming the everyone's right "to the protection of the moral and material interests" resulting from their scientific productions. This equitable sharing of benefits concept is enunciated in CBD which establishes as an international legal principle that the benefits of the use of genetic resources should be equitably shared. Similarly, explicitly in a bioethics context, the UDBHR calls for "equitable access to medical, scientific and technological developments as well as the greatest possible flow and the rapid sharing of knowledge concerning those developments and the sharing of benefits, with particular attention to the needs of developing countries". The FAO International Treaty on Plant Genetic Resources for Food and Agriculture (Seed treaty) establishes a multilateral system of benefit-sharing for the use of plant genetic resources. As a mean of generating benefits from biotechnological research, the IP system and in particular the patent system could have a potential ancillary role in helping to generate, clarify and equitably apportion such benefits. How to find this balance of interests and to determine what is an equitable sharing of benefits, remains controversial? It is another area where formal legal requirements may overlap or be influenced by ethical ideas about what is fair or equitable. This may go beyond the idea of simply apportioning shares of the financial returns. It may also be expressed in terms of providing favourable access to the technology. For instance, PIPRA and some research universities are developing 'humanitarian licensing' measures which provide guarantees of access to life sciences technologies to serve the needs of developing countries; while not legally bound to do so, some follow this policy for ethical reasons. The Public Intellectual Property Resource for Agriculture (PIPRA) initiative has developed licensing language for a humanitarian use reservation of rights, which includes the following: 'University hereby reserves an irrevocable, non-exclusive right in the Invention/Germplasm for Humanitarian Purposes. Such Humanitarian Purposes shall expressly exclude the right for the not-for-profit organization and/or the Developing Country, or any individual or organization therein, to export or sell the Germplasm, seed, propagation materials or crops from the Developing Country into a market outside of the Developing Country where a commercial licensee has introduced or will introduce a product embodying the Invention/Germplasm.'

4.4 Pluralism: Conciliating Different Value Systems

A community's sense of morality and the values of that community may guide ethical judgments. Naturally, these values differ between societies, and the moral basis of ethical judgments will also differ. A technology that is considered immoral in one country may be considered morally acceptable, indeed positively desirable, in another. There has been ethical debate about the patenting of mammals – such as mice bred to be highly susceptible to cancer, for use in medical research. Some argue that patenting genetically modified mammals – however inventive they may be – is inherently immoral. Others take the view that a utilitarian balancing of welfare effects is required. Still others view the question as ethically neutral. Some aspects of stem cell research fall into this category. This raises the question of how the IP system should deal with these different value systems, for instance in the interpretation and application of exceptions in patent law for technology that is contrary to morality.

5. IP and Bioethics

A wide range of ethical questions arises in the debate over biotech IP rights. At times these questions can relate more to the technical field than to the IP right in relation to a certain technology. Ethical judgments may concern choices by the State or by government authorities, or may apply to the behaviour of individuals, firms or institutions. Such issues are not clinically isolated from one another. Nonetheless, given the complexity of issues, it can be helpful to observe some conceptual, legal and ethical distinctions.

Working through the ethical issues can therefore be facilitated by grouping them into four clusters.

6. Ethical Aspects of Technology

This aspect concerns ethical judgments over such matters as forms of research, including research involving human subjects and genetic materials, and over technologies such as genetic engineering. Bioethical issues may arise over such technologies whether or not they are patented. Whether the human body as well as precursors to the human body, simple discovery of parts of the human body, including genes (DNA sequences) and subsequences should be patented? Whether patents on isolated human genes (DNA sequences) should only be obtained, if the patentee has demonstrated a practical and commercial use of the invention? The patents on human genes should not be used to inhibit research and development and prevent free, non-commercial research. Gene patents should not be used to the detriment of the common good or stand in the way of using new diagnosis and treatment methods.

Important bioethical issues, such as prior informed consent, apply to the very practice of research, long before there is any research outcome that may or may not be patented.

Research on stem cells, particularly embryonic stem cells, has raised considerable ethical debate. The question of whether to permit stem-cell research at all is distinct, and may have distinct ethical aspects, from the question of whether the outcomes of such research should be eligible for patent protection. As another example, some have argued that genetic use restriction technologies (GURTs), which prevent farmers from using harvested seeds for future crops, may be unethical, or alternatively should be legally prohibited; others argue that it is a legitimate technology with a valuable commercial role. But such ethical questions are strictly distinct from whether a patent should be granted over such technologies; a patent on a GURT does not entitle its owner actually to practice the technology, and cancelling a patent on the technology equally doesn't prevent its use.

Certain practices may be considered unethical and contrary to morality and consequently directly prohibited. However, such prohibition alone does not automatically prevent the grant of patents related to this knowledge. Not all countries have the same ethical or legal restrictions. Many patent laws exclude explicitly the grant of patents where the exploitation of inventions is considered to be contrary to *ordre public* or morality (such laws are therefore relevant to the following aspect).

On the other hand, biotechnological research in most cases is not merely allowed, but actively encouraged by society, such as the development of new pharmaceuticals. Many technologies have a positive ethical aspect; and some technologies may have ethical and unethical uses. Even pharmaceutical research that is initially promising, but ultimately unsuccessful, may be encouraged and welcomed by society as having a positive ethical character. Nonetheless, it is generally illegal to market a new pharmaceutical without the necessary regulatory approval that follows the successful conclusion of extensive clinical trials. Thus ethical issues in life sciences technology are: i) What life science technologies does society wish to promote and encourage? and ii) What ethical expectations and obligations surround research practices and procedures in the life sciences?

7. Ethical Aspects of Granting Exclusive IP Rights Over a Technology

It is a separate ethical or moral question to consider the kinds of inventions over which national authorities should grant patent rights. As we have noted, some technologies are considered morally desirable (say, new surgical methods), but are still excluded from patent protection in some countries. In other cases, patent protection may be denied in some countries exactly because it would be contrary to morality to commercially exploit the technology. For example, methods of cloning human beings which is patentable in some countries but non-patentable subject matter in most of the countries. In practice, national patent laws typically preclude some forms or categories of technology as being ineligible for patent protection – as being 'unpatentable subject matter.' This question has a long history in patent law, and international negotiators, national legislators, patent authorities, and courts have all been involved in establishing and applying rules in this area. For example, a recurring issue in the application of patent law to biotechnology – dating back a century or more – is how to distinguish a patentable invention from a mere discovery. For instance, when should a patent be granted for a chemical structure when newly isolated from the human body?

Some national legislatures choose to exclude from patent protection classes of inventions that would otherwise be eligible. Such choices can be influenced by ethical concerns, within a broader mix of public policy considerations. For instance, guided by various public policy reasons, including ethical considerations, some countries have chosen to exclude from patentability methods of medical treatment, even when they would otherwise be considered new, inventive and useful. For example, Indian Patents Act, 1970 Sec. 3(i) says 'Any process for the medicinal, surgical, curative, prophylactic, diagnostic, therapeutic or other treatment of animals to render them free of disease or to increase their economic value or that of their products' is non-patentable subject matter. The international WTO-TRIPs rules permit national laws to exclude 'diagnostic, therapeutic and surgical methods for the treatment of humans or animals'. 'This exclusion is based, of course, on a public policy choice not to cover such methods under the patent system, not because of any negative ethical judgment about novel medical treatments which may be of great social benefit. By contrast, other countries choose to allow patents for methods of medical treatment, presumably judging the grant of patents on such methods to have a predominantly positive ethical and policy character.

In parallel with the debate about the ethics of genetically modified organisms, there has been extensive debate about the ethics of patenting life forms, particularly higher life forms such as genetically modified mammals. National policymakers have chosen to resolve these issues in different ways; these differences correspond in part to different ethical perspectives and other social values. Controversy continues to swirl around the patenting of genes or DNA sequences, especially without disclosing any specific known utility. Is it morally sound for society to grant exclusive property rights over nucleotide sequences that are derived from the human genome, when no specific use has been found and disclosed for the patented sequence? Are human genes ethically distinct from any other nucleotide sequences – in general, and especially when patents are concerned? Some have argued, on various policy, legal and ethical grounds, which the bare information provided in human gene sequences should not be patented. Others point to the positive benefits for society of securing clear property rights over useful genes, isolated from their natural setting, to promote the investment of resources in the creation of new forms of diagnosis and therapy. But this debate does not take issue with the ethical aspects of the sequencing of the human genome as such, which has been widely welcomed.

Another choice policymakers and, in individual cases, the patent authorities have to make is to define and apply the concepts of morality and *ordre public* guiding the application of specific exceptions to patentability based on these criteria. WTO-TRIPs agreement says 'Members of the WTO may exclude from patentability inventions, the prevention within their territory of the commercial exploitation of which is necessary to protect *ordre public* or morality, including to protect human, animal or plant life or health or to avoid serious prejudice to the environment, provided that such exclusion is not made merely because the exploitation is prohibited by their law'. For example, the European Biotechnology Directive (98/44/EC) articulates the principle that inventions should be considered unpatentable where their commercial exploitation would be contrary to the *ordre public* or morality. As an example of technology that is unpatentable on the basis of that it is morally unacceptable, it cites "processes for modifying the genetic identity of animals which are likely to cause them suffering without any substantial medical benefit to man or animal, and also animals resulting from such processes". Thus patenting life sciences technology there are some ethical issues:

- What are the ethical questions that need to be weighed when considering whether or not to patent a controversial life sciences technology?

- When should a patent be denied on moral grounds for a life science technology that would otherwise be considered patentable? How should divergent ethical perspectives be reconciled?

- Should the grant of a patent over a certain technology be considered ethically neutral, distinct from regulation of that technology as such, or should it be considered a kind of badge of society's ethical approval of the technology?

8.   Ethical Aspects of Seeking Exclusive IP Rights Over a Technology

We have just reviewed the ethical dimension of the decisions within national legislation to determine that some inventions should be unpatentable. But the actions and decisions of individuals also have an ethical dimension. Thus there may be ethical considerations in the choices of an individual actor – a firm, a research institution, or a university – over whether or not to pursue a patent for a particular invention, even if the invention as such would be legally eligible for a patent under national patent law. Again, there is an uncertain overlap between the legal and ethical aspects.

Thus, some argue that there should be constraints on seeking a patent for an invention which is based on genetic resources or traditional knowledge obtained without prior informed consent and without equitable benefit-sharing. Many cases of biopiracy have been discussed in the chapters on Biodiversity and Traditional Knowledge. This argument can arise even when the claimed invention would otherwise be eligible for patent protection. But are these constraints moral, legal, or both – or should moral constraints harden into legal ones? In fact, legal measures to address these concerns have been introduced in some national laws, and have been proposed for international law. This issue has therefore been dealt with as a legal matter, in terms of both international and national law, but it may also have a continuing ethical aspect. For instance, what if the traditional knowledge that led to an invention was obtained strictly legally – in the sense that no law was broken – but nonetheless the actions of the patent applicant in obtaining and using that knowledge are considered unethical? Some laws have constraints on obtaining or exercising patent rights when they have been secured through inequitable behavior. While this is technically a legal question it may also be considered to have an ethical dimension – what forms of inequitable or improper behaviour should graduate from being considered simply unethical to being illegal, and prevented or punished by legal measures? Does the source of funding or the nature of other inputs to the invention like human tissue samples affect the ethical dimension of applying for a patent?

9. Ethical Aspects of Exercising Exclusive Rights Over a Technology

Some ethical questions do not directly concern the ethics of a technology as such, nor whether that technology should be patented, but rather arise over the ethics of how a patent holder chooses to exercise the rights granted by a patent. In some cases, there may be ethical questions about how a patent holder should exercise his or her patent. Certain ethical constraints could be argued to apply still when a patent holder operates within his or her legal entitlements, while still attracting ethical scrutiny [See Box 1].

Box 1

Breast cancer cell – Myriad Genetics Inc

BRCA-1 and BRCA-2 are two genes linked to susceptibility for breast and ovarian cancer. The risk of falling ill increases if these genes show certain mutations. Identifying the mutations is therefore important for diagnosis and for monitoring higher-risk women. Myriad Genetics Inc., in collaboration with the University of Utah, sequenced the BRCA-1 gene, and applied for patent protection in 1994. The ensuing multifaceted debate over this patent partly concerned the ethical dimension of how a patent on a valuable diagnostic test should be licensed.

How a patent right is licensed becomes important when there is strong public interest in the patented technology. The licensing of patented diagnostic tools has been debated recently. In particular, if a patent holder has exclusive rights over a diagnostic tool, and chooses to license those rights in a restrictive way, could it be argued that there are ethical obligations on that patent holder to grant wider access, even if the licensing approach is strictly legal? Yet we often rely on private capital to carry forward valuable technologies to reach the public: how do differences in the public/private mix of inputs into research and development affect our ethical expectations, as against legal obligations on patent holders?

Public interest safeguards apply to licensing and exploitation of IP rights. Legal restrictions delimitate how IP rights can be exercised in the marketplace. This includes the application of general competition principles, rules against abusive licensing practices and the application of specific remedies under patent law (such as compulsory licensing). But how one chooses to exercise IP rights can also be influenced by the ethical, 'best practice' and policy guidelines on licensing of key technologies. Examples include a growing practice among university technology offices to include model humanitarian provisions in their technology licensing agreements, and a set of OECD Guidelines for the licensing of genetic inventions which suggest a relatively open approach to licensing, particularly for genetic tests.

Access to the benefits of scientific research may also be developed through other licensing structures. For instance the BiOS initiative characterizes open source licensing in the life sciences as follows:

"Usually, licenses for patented technology impose strict conditions on the user, commonly involving fees or royalties for use of the materials or methods or both. Material Transfer Agreements (MTAs) typically impose the condition that the technology may only be used for certain purposes, often not allowing the development of products. Instead of royalties or other conditions that disfavor creation of products, under a BiOS-compliant agreement, in order to obtain the right to use the technology, the user must agree to conditions that encourage cooperation and development of the technology.

These conditions are that licensees cannot appropriate the fundamental "kernel" of the technology and improvements exclusively for themselves. The base technology remains the property of whatever entity developed it, but improvements can be shared with others that support the development of a protected commons around the technology, and all those who agree to the same terms of sharing obtain access to improvements, and other information, such as regulatory and biosafety data, shared by others who have agreed.

In other words, to maintain legal access to the technology, you must agree not to prevent others who have agreed to the same terms from using the technology and any improvements in the development of different products." Thus some ethical issues on Exercising exclusive rights over life sciences technology are:

- What are the strictly ethical factors – apart from legal regulation – that should be weighed in determining how best to exercise the rights under a patent on a new life sciences technology?

- How does the ethical dimension of exercising a patent change according to factors, such as the degree of public sector funding, or the contribution of human genetic samples, to research activities? Is it different for public and private sector players? How?

- Does the humanitarian value of a patented technology affect the ethical dimension of how the patent should be exercised or licensed?

Selected reading

http://www.wipo.int/Intellectual Property and Bioethics – An Overview

Ninawe, A.S. Intellectual property rights in biotechnology. *In*: Protecting intellectual property in life sciences (Eds. Dominic Keating, Abha Agnihotri and Ajit Varma), Amity University Press, Noida, India, pp. 152-161.

Index